WAITING FOR *TSUNAMI*
Coastal Hazards of Northern San Diego County, California

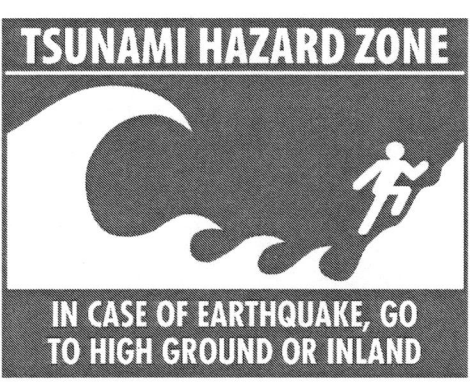

Field trip participants on Scripps Pier (Stop 4). September 8, 2012.

Back row: *Joe Corones, Neal Driscoll (Scripps Pier host), Diane Murbach, Mike Hart, Cheryl Peach (Scripps Pier host), Al Trujillo, Ric Siem, Daniel Pankratz, Don Barrie, Chris Metzler, Gina Bochicchio, Greg Farrand, Jeff Miller, Brenden Hawk, Tom Devine, Dave Bloom, Norrie Robbins, Germar Bernhard, Anne Hoppe, Phil Farquharson, Andrew Pigniolo, Lawrence Busch, Todd Wirths, Ray Rector, John Turbeville, Rob Hawk, Monique Hawk, Stephen Jacobs, Donna Gooley.*

Front row: *Margaret Eggers, Jen Morton, Nissa Morton, Lowell Lindsay, Diana Lindsay, Bryan Miller-Hicks, Brenda Hawk, Mary Walke, Patty Deen, Katy Freese, Bulent Bas, Sue Tanges, Bert Vogler.*

Field Trip Attendees (not pictured): *Greg Cranham, Cari Gomes, Dan Gomes, Monte Murbach.*

Photo: Monte Murbach.

WAITING FOR *TSUNAMI*
Coastal Hazards of Northern San Diego County, California

*Published in conjunction with the
2012 Field Trip of the
San Diego Association of Geologists
September 8-9, 2012*

Cari Gomes, Editor
Dave Bloom, Project Manager

SAN DIEGO ASSOCIATION OF GEOLOGISTS

Distributed by Sunbelt Publications
www.sunbeltbooks.com

Waiting for Tsunami: Coastal Hazards of Northern San Diego County, California

Copyright © 2012 San Diego Association of Geologists,
a program of San Diego Geological Society, Inc.

Individual papers, photos, and artwork copyright by authors
and used with permission. All rights reserved
First edition 2012, second printing 2012

Edited by Cari Gomes
Project Management by David M. Bloom
Cover concept by Philip T. Farquharson
Book Design by Michael Schrauzer
Printed in United States of America

San Diego Association of Geologists
P.O. Box 191126
San Diego, California 92159-1126
www.sandiegogeologists.org

or the distributor:

Sunbelt Publications
1256 Fayette Street
El Cajon, California 92020
(619) 258-4911, fax: (619) 258-4916

15 14 13 12   5 4 3 2

Library of Congress Cataloging-in-Publication Data

Waiting for tsunami : coastal hazards of northern San Diego County,
California / Cari Gomes, editor. -- 1st ed.
    p. cm.
  ISBN 978-0-916251-23-9 (alk. paper)
  1. Tsunamis--Risk assessment--California--San Diego County. 2.
Tsunamis--California--San Diego County--History. 3. Coasts--California--San
Diego County. 4. Geology--California--San Diego County. I. Gomes, Cari.
  GC222.C2W35 2012
  363.34'942--dc23
                    2012039519

The San Diego Geological Society, Inc., gratefully acknowledges the generous contributions of our sponsors, including Diamond Sponsors H & P Mobile Geochemistry, Inc., and Hargis + Associates, Inc., to the publication of this book.

Front Cover: Woodrow L. Higdon, Geo-Tech Imagery
Inside Front Cover: Map adapted from Ward (2002)
Inside Back Cover: Geological Map, modified from Blake (1855)
Back Cover: Map adapted from Sheet 1, Dartnell and others (2007)

Table of Contents

Preface, *Cari Gomes* vii
Introduction viii
Acknowledgments x
Dedication xi
Sponsors xii
Officers xiii
SDAG Publications xiv
Related Publications xvi
Geologic Time Scale xvii
List of Abbreviations xviii
Tsunami Inundation Map xix
Campground Vicinity Map xx

I. ROAD LOG

Waiting for Tsunami: Coastal Hazards of Northern San Diego County, California, Cari Gomes, Field Trip Leader
Dave Bloom 1

Day 1: Field Trip Stops 1 through 4 2
    Route Map Showing Field Trip Stops 1 through 4
    Sidebar: Geologic History of San Diego, *Richard P. Phillips*
    Stop 1. Rancho Guajome to Morro Hill
    Stop 2. Morro Hill to San Onofre Nuclear Power Plant
    Stop 3. San Onofre State Beach
    Stop 4. San Onofre to Scripps Institution of Oceanography Pier
    Sidebar: Chilean Quake Tsunami Caused Damages
        in San Diego Harbor, *John Campbell*
    Sidebar: History of Guajome Regional Park
        Land and People, *Eleanora I. Robbins*

Day 2: Field Trip Stops 5 through 7 29
    Route Map Showing Field Trip Stops 5 through 7
    Stop 5. Rancho Guajome to Cerro de la Calavera
    Stop 6. Calavera to Agua Hedionda
    Stop 7. Agua Hedionda to Beacons Beach

CONTENTS

II. PAPERS 55

Preliminary Search for Eltanin Impact Tsunami
Deposits in Southern California:
A Possible Early Pleistocene Chronozone
*Lawrence L. Busch, Brian J. Swanson,
Janis L. Hernandez, and Brian P.E. Olson* 57

Preliminary Analysis of Piedra de Lumbre and
Talega Canyon Cherts: Distinctive and Historically Significant
Outcrops on Camp Pendleton,
San Diego County, California
*Eleanora I. Robbins, Andrew R. Pigniolo,
Greg T. Cranham, and William J. Elliott* 83

Lawrence Canyon Fault, Oceanside, California
William J. Elliott and Dr. Monte Marshall 119

Landslide Gaps at Beacon's Beach, Encinitas, California
William J. Elliott 137

Shoreline History and Mystery, Solana Beach, California
William J. Elliott 141

A Case for the Clean Sand Layer within
the Bay Point Formation in Solana Beach
*Walter F. Crampton, James Knowlton,
Gregory A. Spaulding, and Braven R. Smillie* 159

Tsunami History of San Diego
Duncan Carr Agnew 173

Theoretical Aspects of Tsunamis
along the San Diego Coastline
W. G. Van Dorn 179

The Rose Canyon Fault Zone in San Diego
Thomas Rockwell 183

Spelunking on San Diego's Coastline
Gregory A. Spaulding and Walter F. Crampton 203

Archaeological Investigations of the Sunken Gardens of
Mission San Luis Rey, California
Jack S. Williams and Anita G. Cohen-Williams 231

PREFACE

Welcome to the San Diego Association of Geologists (SDAG) 2012 annual field trip! This guidebook, *Waiting for Tsunami: Coastal Hazards of Northern San Diego County*, is the field guide that accompanies the trip.

Being born and raised in Hilo, Hawai'i, dubbed the Tsunami Capital of the World, I have always had a special awareness about tsunamis as geologic hazards. For me and other residents of Hilo Town, tsunamis are a way of life. Periodically, school would be canceled during tsunami warnings, which we called "tsunami days," the Hilo equivalent of snow days in certain northern areas of the U.S. mainland. It was not until after I moved to the North American continent that I realized the topic of tsunamis was not at the forefront of discussions about our nation's geologic hazards. Although some tsunami assessment studies have been completed for West Coast locations, the 2011 Tohoku Earthquake and subsequent tsunami in Japan have only recently brought the potential loss of life and damage due to these killer waves to the attention of coastal residents and planners, particularly in California.

The 2012 SDAG field trip begins on Saturday September 8 at the San Onofre Nuclear Generating Station (SONGS). We visit the facility and discuss the benefits of nuclear power generation as well as the potential risks to the plant associated with tsunami and tectonic activity in the area. We also explore the landslides near the structure, where we discuss the factors that influence the size, shape, and dangers of these features. The first day of the trip ends with a tour of Scripps Pier in La Jolla and the area's associated geologic features. Sunday September 9 begins with a stop at Cerro de la Calavera in Carlsbad where we explore the history of this Miocene volcanic plug. Next, we head to the coast to see Agua Hedionda Lagoon, Carlsbad, where we examine outcrops that have been interpreted as tsunami deposits. We conclude our adventure at Beacons landslide.

Cari Gomes
SDAG 2012 Field Trip Chair

INTRODUCTION

TSUNAMIS

The world has a long history of tsunamis that has caused death and destruction throughout human history. As early as 426 BC, the Greek historian Thucydides inquired in his book *History of the Peloponnesian War* about the causes of tsunami, and was the first to argue that ocean earthquakes must be a factor in generating these waves. The historic 1883 eruption of Krakatau volcano in Indonesia generated a tsunami greater than 35 meters (115 feet) high that drowned more than 1000 villages and killing more than 36,000 people. The 1946 earthquake that shook Scotch Cap, Alaska, created a tsunami that funneled through Hilo Bay to generate a surge of water almost 17 meters (56 feet) above the normal high tide, causing $25 million in damages and killing 159 people nearly 3000 miles away.

Tsunami evacuation route, Ocean Beach, 2012. Photo: Diane Murbach.

Japan has a long recorded history of tsunami (*tsu*= harbor, *nami*= waves). Most recently, Japan's 2011 magnitude 9.0 Tohoku earthquake generated a tsunami that reached a record height of 40 meters (131 feet) in one location because of amplification by offshore topography. The tsunami overtopped many protective seawalls and killed at least 19,500 people. In all, it caused damages estimated at $235 billion, making it the most expensive natural disaster in world history.

Tsunamis typically originate from sudden movements of the sea floor such as turbidity currents, fault movements with vertical displacements, or underwater volcanic eruptions. In rarer cases, tsunamis may also be caused by landslides or meteorite or asteroid impacts. Regardless of the mechanism that generates the event, the sudden sea floor movement causes a displacement of the overlying water within the water column. This displacement generates a series of waves, typically around 200 kilometers (125 miles) long that move in excess of 700 kilometers (435 miles) per hour in the open ocean.

When these waves encounter shallower coastal water, the period of the waves remains the same, but their velocity is reduced, causing the waves to stack up, creating a greatly increased wave height. This results in a series of surges and withdrawals of water at the coast lasting several hours. In some cases, the crest of the tsunami arrives first, resulting in a huge amount of seawater rapidly rushing ashore; in other cases, the trough arrives first, and the coastal region experiences a huge withdrawal of seawater.

OTHER COASTAL HAZARDS: MASS WASTING

Population studies suggest that roughly 80% of California's population lives within 50 kilometers (31 miles) of the coast. California's heavily populated coastal regions are mostly underlain by soft sedimentary and metamorphic rock that is actively eroding. This erosion occurs through one of three processes: (1) wave attack, (2) mass movements, and (3) seismic shaking. The erosion rates of the relatively weak rock typical of Southern California can be fairly high (as much as 60 centimeters [24 inches] per year) but highly episodic. California's large coastal population coupled with actively eroding shorelines presents many potential hazards for residents.

The northern San Diego County coastline has been subject to many large landslide events—some current and some prehistoric. Failure modes include block glides along bedding planes that dip seaward and smaller topples and slab failures that occur as a result of wave-cut notching at the base of coastal bluffs. Primary areas of the San Diego coast where large block glide landslides have occurred include San Onofre (Echo Arch), Encinitas, Torrey Pines State Park, Black's Beach, and Leucadia. We examine some of these landslide features during this field trip.

Warning sign posted by the City of Encinitas in a parking lot overlooking the beach, April 2012. Photo: Cari Gomes.

ACKNOWLEDGMENTS

Field Trip Committee
Dave Bloom
Greg Cranham
Phil Farquharson
Diane Murbach
Monte Murbach

Field Trip Stop Leaders
Neal Driscoll
Phil Farquharson
Tom Freeman
Mike Hart
Phil Rosenberg
Dave Schug

Cover Sponsors
H&P Mobile Geochemistry, Inc.
Hargis + Associates, Inc.

Professional
Margaret Eggers
Bill Elliott
Lowell Lindsay
Anne Sturz
Al Trujillo
Carole Ziegler

Personal
Finally, I would like to thank my husband, Dan Gomes, whose love and support make anything possible.

Yours Truly,
Cari Gomes

DEDICATION TO
MICHAEL W. "MIKE" HART
AKA "MR. LANDSLIDE"

Mike Hart Mapping fissures on the Mt. Soledad Landslide, October 2007

When the 2012 field trip planning committee thought about whom to dedicate this guidebook, there was only one choice: Mike Hart, also known as Mr. Landslide! Mike will lead field trip attendees to landslides at San Onofre State Beach.

Mike has been active in the geologic community since 1970. Mike is a Certified Engineering Geologist in California and has practiced in the southern California area for 40 years. He is a graduate of San Diego State University (M.S., 1972). In 1975 he joined Geocon, Inc., becoming vice president and principal engineering geologist. Since leaving Geocon in 1992 he has been an independent engineering geology consultant specializing in landsliding and evaluation of fault hazards.

Recent publications include papers on landslide failure mechanisms, and identification. Mike is past chairman of the Association of Environmental and Engineering Geologists Landslide Committee, past Chair of the Engineering Geology Division of the Geological Society of America, and a long time member and past president of the San Diego Association of Geologists (1973) and South Coast Geological Society (2003). Mike currently serves on the Board of Directors for SDAG's non-profit San Diego Geological Society, Inc. (SDGS) established in 2008.

SDAG 2012 FIELD TRIP COMMITTEE:
*Cari Gomes, Dave Bloom, Greg Cranham,
Monte & Diane Murbach, Lowell Lindsay, Phil Farquharson.*

Sponsors

Diamond ($1000)

H & P Mobile Geochemistry, Inc.
www.handpmg.com

Hargis + Associates, Inc.
www.hargis.com

Ruby ($500)

Jim Ashby, Mission Geoscience
Steve Zigan
Dave Bloom/Tetra Tech
Dan Chambers, Chambers Environmental

Emerald ($100)

Dr. Pat Abbott
Sally & Dennis Avery
Richard W. Berry
Curtis Burdett
Joe Corones
Greg Cranham
Cheryl and Neal Driscoll
Bill Elliott
Eggers Environmental, Inc.
Greg Farrand
Phil Farquharson, CG-Squared Productions
Don Wilson Clark/Katy Freese
Carolyn Glockhoff, Caro-Lion Enterprises
Cari Gomes
Sarah Gray
Rob Hawk
Jonas & Associates
Diana & Lowell Lindsay, Sunbelt Publications
Ninyo & Moore
Monte Marshall
Monte & Diane Murbach, Murbach Geotech
Brian Olson
Peet's Coffee
Les Reed, Geotechnical Exploration Inc. (GEI)
Rigid Lifelines
San Marcos Brewery
SEALASKA Environmental Services LLC, Doug Peeler
Gerald Shiller
Anne Sturz
Sue Tanges, Southland Geotechnical Consultants
Southwest Geophysics, Inc.
TerraCosta Consulting Group, Inc.
Bob Smillie, TerraCosta Consulting Group, Inc.
David & Jan Steller
Mary Walke
youtubedownloadersite.com
Carole Ziegler

The generosity of these sponsors allows our organization to keep membership dues low, reduce rates for students, produce monthly newsletters, purchase supplies, host the annual field trip, produce our geologic guidebooks, and provide scholarships to local geology students. To contact any of the above sponsors, please visit www.sandiegogeologists.org.

Officers

San Diego Association of Geologists

PRESIDENT
Todd Wirths

VICE PRESIDENT
Cari Gomes

SECRETARY
Brian Olson

TREASURER
Jennifer Morton

PUBLICATIONS
Lowell Lindsay

WEBMASTER
Carolyn Glockoff

San Diego Geological Society, Inc.

PRESIDENT
Dave Bloom

VICE PRESIDENT
Mike Hart

SECRETARY
Diane Murbach

TREASURER
Greg Cranham

DIRECTOR
Lowell Lindsay

DIRECTOR
Monte Murbach

DIRECTOR (EX OFFICIO)
Todd Wirths

San Diego Association of Geologists is a program of San Diego Geological Society, Inc., a California 501(c)(3) nonprofit corporation.

San Diego Association of Geologists Publications

1972 *Field Trip Guide, Stratigraphy and Structure Southeast of San Diego, California*. Ernest Artim, ed.

1973 *Studies on the Geology and Geologic Hazards of the Greater San Diego Area, California*, Arnold Ross and Robert J. Dowlen, eds.

1974 *Recent Geological and Hydrologic Studies, Eastern San Diego County and Adjacent Areas: a Guidebook*, Michael W. Hart and Robert J. Dowlen, eds.

1975 *Studies on the Geology of Camp Pendleton, and Western San Diego County, California*, Arnold Ross and Robert J. Dowlen, eds.

1977 *Geology of Southwestern San Diego County and Northwestern Baja California*, Gregory T. Farrand, ed.

1978 *Natural History of the Coronado Islands, Baja California, Mexico*, Herman T. Kuper, ed.

1979 *Earthquakes and other Perils, San Diego Region*, Patrick L. Abbott and William J. Elliott, eds. *(GSA annual meeting field trip guidebook).*

1981 *Geologic Investigations of the Coastal Plain, San Diego County, California*, Patrick L. Abbott and Shannon O'Dunn, eds.

1982 *Geologic Studies in San Diego*, Patrick L. Abbott, ed.

1985 *On the manner of Deposition of Eocene Strata in Northern San Diego County*, Patrick L. Abbott, ed.

1987 *Selected Geological and Historical Aspects of the Julian Gold Mining District*, George Copenhaver and Ron Kofron, eds.

1988 *Landslides in Crystalline Basement Terrain*, James R. Evans, ed.

1989 *The Seismic Risk in the San Diego Region: Special Focus on the Rose Canyon Fault System, Proceedings & SDAG Field Trip*, Glenn Roquemore, Susan Tanges, Marion Wright, Michael Reichle, Thomas Heaton, Diane Murbach, and Gilbert Najera, eds.

1990 *Geotechnical Engineering Case Histories in San Diego County*, John H. Hoobs, ed.

1991 *Environmental Perils, San Diego Region*, Patrick L. Abbott and William J. Elliott, eds. *(GSA annual meeting field trip guidebook).*

SAN DIEGO ASSOCIATION OF GEOLOGISTS PUBLICATIONS

1992 *Natural History of the Coronado Islands, Baja California, Mexico (revisited 1992)*, Lynn Perry, ed.

1993 *Colorado Desert and Salton Trough Geology*, Joe Corones, ed.

1994 *Geology and Natural History of Camp Pendleton, U.S.M.C. Base, San Diego County, California*, Phillip S. Rosenberg, ed. *(2nd Edition, 2010)*.

1995 *Paleontology and Geology of the Western Salton Trough Detachment, Anza-Borrego Desert State Park, California*, Paul Remeika and Anne Sturz, eds.

1996 *Geology and Natural Resources of Coastal San Diego County, California*, Tissa Munasinghe and Phillip Rosenberg, eds.

1997 *Santa Cruz Island: Geology Field Trip Guide*, James R. Boles and Werner Landry, eds.

1998 *Geology and Geothermal Resources of the Imperial and Mexicali Valleys*, Lowell E. Lindsay and William Hample, eds.

1999 *Water for Southern California: Water Resources Development at the Close of the Century*, Gregory T. Cranham, ed.

2000 *Geology and Enology of the Temecula Valley, Riverside County, California*, Barbara B. Birnbaum and Kerry D. Cato, eds.

2001 *Coastal Processes and Engineering Geology of San Diego, California*, Robert C. Stroh, ed.

2003 *Geology of the Elsinore Fault Zone, San Diego Region*, Monte L. Murbach and Michael W. Hart, eds. *(joint publication with SCGS)*

2004 *Mining History and Geology of Joshua Tree National Park*, Margaret R. Eggers, ed.

2005/06 *Geology and History of Southeastern San Diego County, California*, David M. Bloom, Philip T. Farquharson, and Carole L. Ziegler, eds.

2006 *Mines and Geology of the Randsburg Area, An Historical Gem of the Mojave Desert*, by D.D. Trent.

2010 *Geology and Lore of the Northern Anza-Borrego Desert Region*, Monte L. Murbach and Charles C. Houser, eds. *(joint publication with SCGS)*.

2011 *Picacho and the Cargo Muchachos*, Todd Wirths, ed.

2012 *Roadside Geology along Sunrise Highway*, by Michael J. Walawender.

Related Publications

Publications describing the general geology, coastal environment, and regional landsliding include:

Sea Cliffs, Beaches, and Coastal Valleys of San Diego County: Some Amazing Histories and Some Horrifying Implications (Kuhn and Shepard, 1984, University of California Press).

The Regressive Pleistocene Shorelines, Coastal Southern California (Heath and Lewis eds., 1992, Guide Book No. 20, South Coast Geological Society).

Geologic Guide of San Onofre Nuclear Generating Station and Adjacent Regions of Southern California (Fife ed., 1979, Guide Book 46, Pacific Sections of the American Association of Petroleum Geologists, Society of Economic Paleontologists and Mineralogists, and Association of Engineering Geologists).

Guidebook for Field Trips, San Diego County, 1961, Annual Meeting of the Cordilleran Section of the Geological Society of America, 57th, San Diego State College.

The Rise and Fall of San Diego: 150 Million Years of History Recorded in Sedimentary Rocks (Patrick Abbott, 1999, Sunbelt Publications).

Geologic hazards in San Diego: earthquakes, landslides, and floods (San Diego Society of Natural History, 1977)

GEOLOGIC TIME SCALE

EON	ERA	PERIOD/SYSTEM		EPOCH/SERIES	Age estimates of boundaries in mega-annum (Ma) unless otherwise noted
PHANEROZOIC	Cenozoic	Quaternary (Q)		Holocene	11,700 ±99 yr
				Pleistocene	2.588
		Tertiary (T)	Neogene	Pliocene	5.332 ±0.005
				Miocene	23.03 ±0.05
			Paleogene	Oligocene	33.9 ±0.1
				Eocene	55.8 ±0.2
				Paleocene	65.5 ±0.3
	Mesozoic	Cretaceous (K)		Late	99.6 ±0.9
				Early	145.5 ±4.0
		Jurassic (J)		Late	161.2 ±4.0
				Middle	175.6 ±2.0
				Early	199.6 ±0.6
		Triassic (TR)		Late	228.7 ±2.0
				Middle	245.0 ±1.5
				Early	251.0 ±0.4
	Paleozoic	Permian		Lopingian / Guadalupian / Cisuralian	299.0 ±0.8
		Carboniferous	Pennsylvanian	Late / Middle / Early	318.1 ±1.3
			Mississippian	Late / Middle / Early	359.2 ±2.5
		Devonian		Late / Middle / Early	416.0 ±2.8
		Silurian		Pridoli / Ludlow / Wenlock / Llandovery	443.7 ±1.5
		Ordovician		Late / Middle / Early	488.3 ±1.7
		Cambrian		Late / Middle / Early	542.0 ±1.0
Precambrian					~4600

Geologic Time Scale.
Reference: U.S. Geological Survey Geologic Names Committee, 2010, Divisions of geologic time—major chronostratigraphic and geochronologic units: U.S. Geological Survey Fact Sheet 2010–3059, 2 p.

LIST OF ABBREVIATIONS AND SYMBOLS

Ave	Avenue
BC	year before Christ
Blvd	Boulevard
CGS	California Geological Survey
cm	centimeter
CO_2	carbon dioxide
et al.	*et alia* (latin for "and others")
g/cc	grams per cubic centimeter
ka	kiloannum, or thousand years ago
Kgd	Cretaceous granodiorite
km	kilometers
kph	kilometers per hour
Kt	Cretaceous tonalite
m	meter
Ma	million years ago
MCB	Marine Corps Base
mm	millimeter
M_s	Earthquake surface wave magnitude
MSL	mean sea level
M_w	moment magnitude, successor to Richter magnitude
M.y.	million years
mya	million years ago
NRC	Nuclear Regulatory Commission
NOAA	National Oceanographic and Atmospheric Administration
PDT	Pacific Daylight Time
QTs	sandstone (Quaternary or Tertiary)
RBSP	Regional Beach Sand Project
SANDAG	San Diego Association of Governments
SCE	Southern California Edison
SDAG	San Diego Association of Geologists
SDGS	San Diego Geological Society
SIO	Scripps Institution of Oceanography
SONGS	San Onofre Nuclear Generating Station
SWFSC	Southwest Fisheries Science Center
Tda	Tertiary dacite rocks (volcanic plug)
Ti(r)	Tertiary intrusive (hypabyssal)-rhyolite
Tsa	Santiago Formation
Tv	Tertiary volcanic rocks
Tv(r)	Tertiary volcanic-rhyolite
UC	University of California
UCSD	University of California, San Diego
USGS	United States Geological Survey
UTC	Coordinated Universal Time, a successor to Greenwich Mean Time
yr	year

TSUNAMI INUNDATION MAP

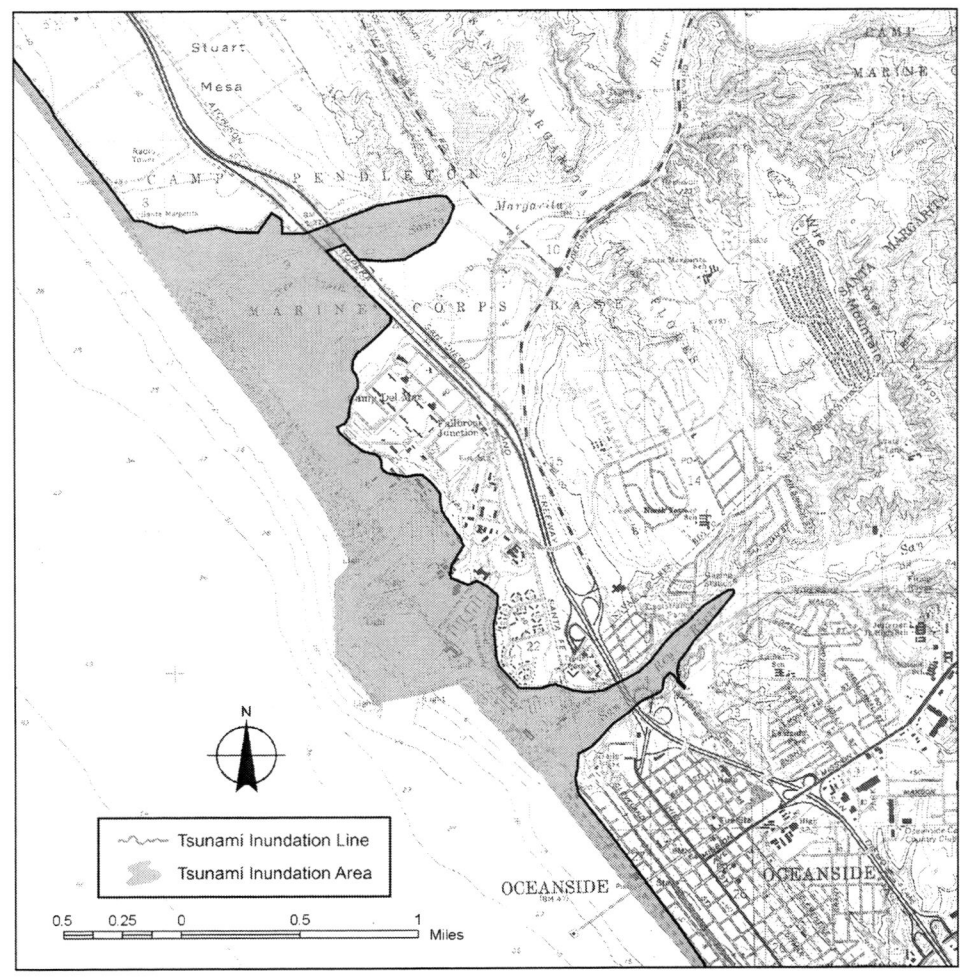

TSUNAMI INUNDATION MAP FOR EMERGENCY PLANNING
OCEANSIDE QUADRANGLE – SAN LUIS REY QUADRANGLE, JUNE 1, 2009
http://www.conservation.ca.gov/cgs/geologic_hazards/Tsunami/Inundation_Maps/SanDiego/Pages/SanDiego.aspx

This tsunami inundation map was prepared to assist cities and counties in identifying their tsunami hazard. It is intended for local jurisdictional, coastal evacuation planning uses only. This map, and the information presented herein, is not a legal document and does not meet disclosure requirements for real estate transactions nor for any other regulatory purpose.

The inundation map has been compiled with best currently available scientific information. The inundation line inland of the shaded arearepresents the maximum considered tsunami runup from a number of extreme, yet realistic, tsunami sources. Tsunamis are rare events; due to a lack of known occurrences in the historical record, this map includes no information about the probability of any tsunami affecting any area within a specific period of time.

State of California Emergency Management Agency, Earthquake and Tsunami Program:
http://www.oes.ca.gov/WebPage/oeswebsite.nsf/Content/B1EC51BA215931768825741F005E8D80?OpenDocument

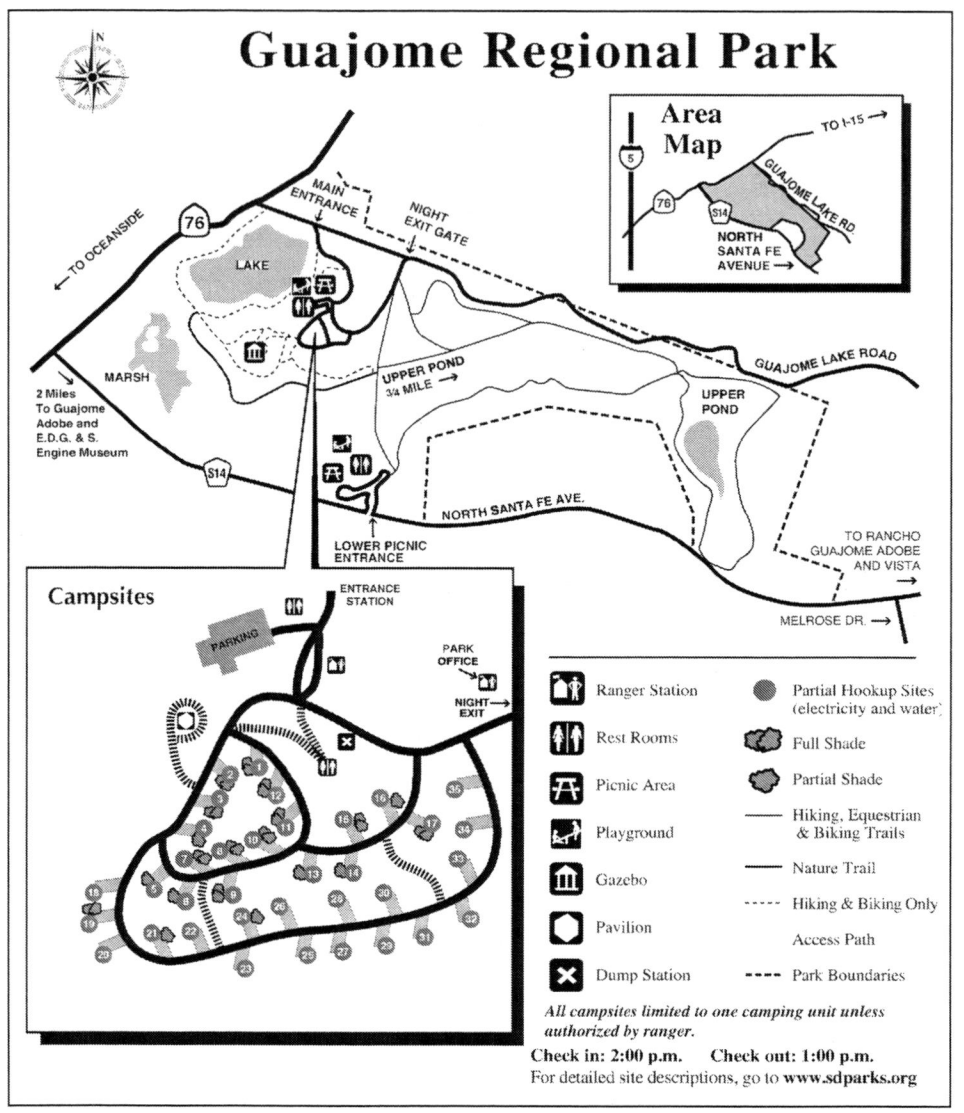

Campground Vicinity Map.
Reference: http://www.sdcounty.ca.gov/reusable_components/ images/parks/doc/GuajomeBrochure.pdf

I. ROAD LOG

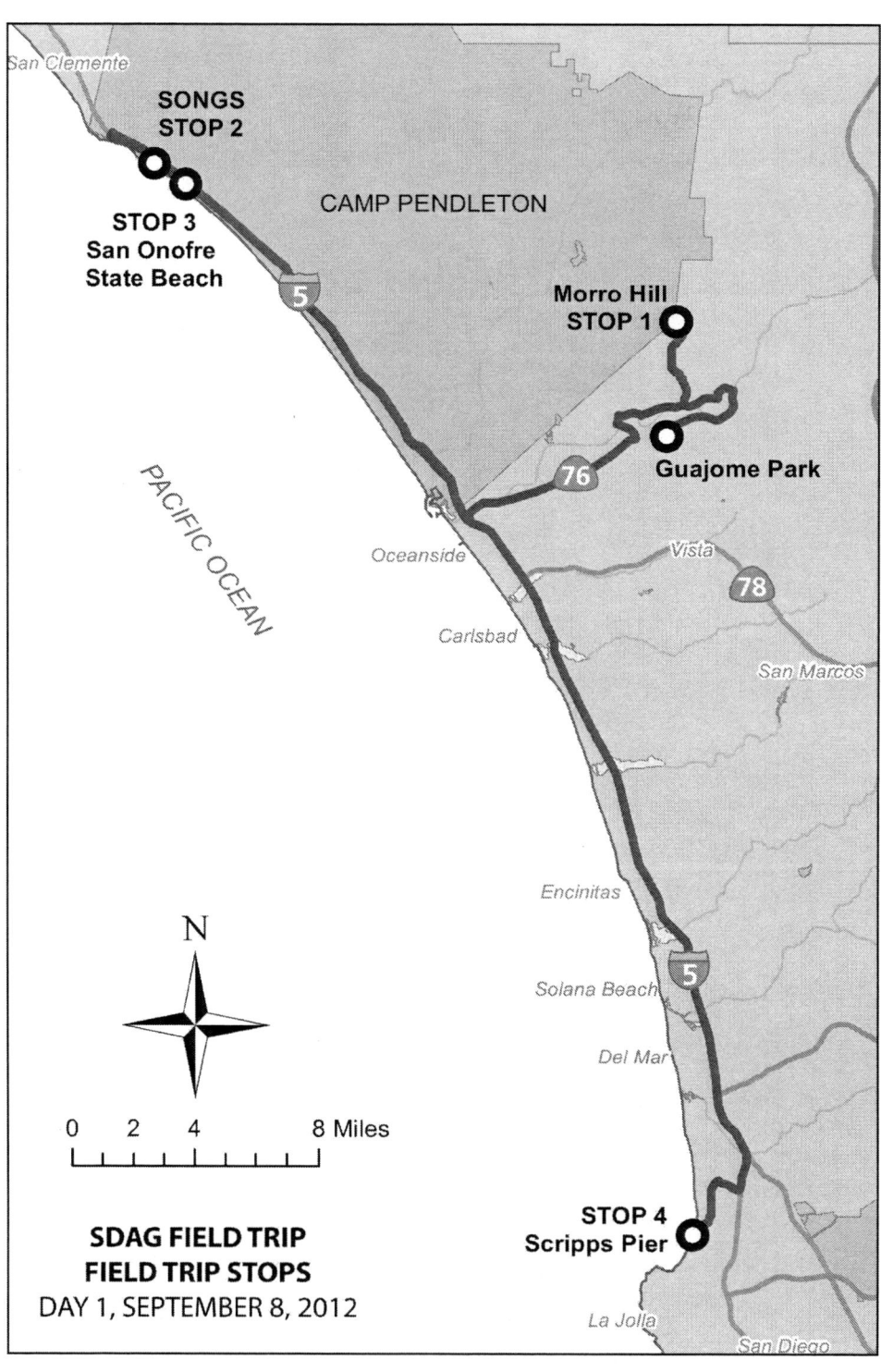

SAN DIEGO ASSOCIATION OF GEOLOGISTS 2012 FIELD TRIP ROAD LOG

Waiting for Tsunami: Coastal Hazards of Northern San Diego County, California

Dave Bloom
dave@cactolith.com

SATURDAY, SEPTEMBER 8, 2012

The trip begins at Guajome Regional Park and campground on the San Luis Rey River straddling the cities of Oceanside and Vista, California. We'll discuss the history and setting of Rancho Guajome when we return at the end of the day. Travel is by chartered bus today.

We will make four stops today. Discussions will cover general geology, outcrops, stratigraphy, environmental issues, history and culture along the way. We will loop north to volcanic Morro Hill for Stop 1 and return back to San Luis Rey River. Our route turns northwest along the nearly undeveloped Pacific coastline of Marine Corps Base Camp Pendleton to Stop 2 at San Onofre Nuclear Generating Station (SONGS). Stop 3 is a short distance south for a view of landslides and a good exposure of the Cristianitos fault. Stop 4 takes us south to Scripps Pier in La Jolla before returning to camp at Guajome.

Leave our campground at Guajome Regional Park. Reset odometer at the park entrance on Guajome Lakes Road.

Odometer (Trip) Mileage

0.0 Exit Guajome Park. Turn left on Guajome Lakes Road.

0.1 Turn right, heading east on Highway 76 (San Luis Rey Mission Expressway).

2.9 Bridge over San Luis Rey River from south to north banks. A second parallel highway bridge spanning the river is nearly complete which will allow four lanes of traffic.

The 55-mile long San Luis Rey River drains a 557 square mile watershed. The river is usually dry having been extensively dammed, diverted for irrigation and domestic consumption. Henshaw Dam, constructed in the upper basin in 1923 by the San Diego County Water Company, provided a source of reliable water coinciding with the water shortage crisis as avocado and citrus agriculture rapidly grew. Following the first arrival of water on February 27, 1926, various crops were planted in increasing numbers, and the Vista area became known as the "Avocado Capital of the World." Vista Irrigation District purchased the San Diego County Water Company in 1946, and currently operates the facility (USGS, 2009; Vista Irrigation District, 2012).

Torrential rains have occasionally caused destructive flooding, for example in 1862, 1884–91, 1927, 1938, 1941, and mostly recently in January 1993. The Army Corps of Engineers built levees in phases beginning in 1987 and completing in 2000, creating a 400 foot wide channel along the lower 7 miles of the river to manage floodwaters (City of Oceanside, 2010). A bike path tops the south levee. Maximum flow of the river is reportedly 25,700 cubic feet per second (USGS, 2009).

3.1 Turn left onto North River Road

5.0 Turn right on Sleeping Indian Road. The sign is small so pay attention.

7.9 Wilshire Road. The high area ahead of you and to the right and left is informally called Sleeping Indian Hill by locals. Bill Elliott (1985) reports four outliers of "Indian Hill volcanics," unconformably overlying Santiago Formation – Member A of Wilson (1972), which he describes as "approximately 8 to 30 meters of thinly- to massively-bedded, well-indurated, unfossiliferous, olive-green claystone, tan to gray siltstone, and light gray to white sandstone," which in turn are nonconformable with underlying Bonsall tonalite. Elliott describes geomorphic features and potential increased irrigation on weak clay soils that could result in reactivation of landslides in this area.

8.1 Morro Hill is on your left. Just before road summit, turn left onto the gravel shoulder where there is a short dirt road. Keep to the right and park at the base of the hill. There is only enough space for about six cars to park, so please carpool. Do not park in driveways of private residents.

GEOLOGIC HISTORY OF THE SAN DIEGO REGION
*Dr. Richard P. Phillips**

Geology has been described as a four-dimensional science. We are concerned not only with the land surface but must look at depth and time as well.

The concept of geologic time often is hard to grasp. The basic idea is that there is lots of it. For example, current age determinations place the earth at 4.6 billion years. Processes that proceed with incredible slowness in terms of human experience can produce profound changes over geologic time. Each rock that we see at the surface of the earth is a product of these slow changes and has a complex history. We have to consider both how it formed and how it reached the surface, two distinct processes.

Geologically speaking, San Diego County has a relatively short history. The oldest rocks we can recognize are from the middle of the Mesozoic Era, the age of dinosaurs. At that time, about 150 million years ago, San Diego was quite possibly an island arc similar to the Japanese islands of today. To the west was a deep sea trench and to the east a shallow sea. Volcanoes were depositing a great pile of andesitic rock from flows and explosive events. This setting is typical of a subduction zone, where one plate of the earth's lithosphere is sliding beneath another. As it descends into the earth's upper mantle, it partially melts, along with some of the upper mantle that lies above it, producing molten rock that rises up through the crust of the overlying plate to form volcanoes.

Not all of that molten rock reached the surface. In fact, most of it remained in the cooler rocks of the overlying crust and crystallized. This molten rock profoundly altered the older rocks. The magma mixed with them and sometimes partially melted them.

Erosion has exposed these older rocks as the metamorphic rocks of San Diego County. The magma that crystallized some 100 million years ago is exposed today as the granitic rock of the southern California batholith, which formed the bulk of the Peninsular Ranges.

By the end of the Mesozoic Era (65 million years ago), geologic quiet prevailed. Erosion worked on the earlier mountains and wore them away. Great thicknesses of sediment were laid down in the sea to the west, which supported a wide variety of invertebrate life. Fossiliferous sedimentary rocks of that time period are exposed today in eastern Carlsbad where dinosaur fossils were recovered, foothills of north-central Camp Pendleton, and sea cliffs of La Jolla and Point Loma.

Erosion continued, and the mountains were worn away completely or buried by their own sediments. Rivers that arose in Sonora, Mexico, flowed westward and deposited great deltas of sand and gravel along the Pacific coast 40 to 50 million years ago. At that time, San Diego was a broad coastal plain, much like the Gulf of Mexico seaboard of today, with a few hills rising above the general level. Santa Margarita and San Marcos mountains likely stood high above the general land surface. Coastal lagoons, rich in oysters and fringed with mangroves and nut palms, bordered the deltas. Rocks of that time (the Eocene Epoch) are being quarried in Carroll Canyon for gravel and, until recently, in Oceanside for silica sand.

During that long period of erosion, a thick blanket of soil formed on the underlying rock. Most of that soil has been removed by subsequent erosion. Only in those few places is it preserved. In many places in San Diego County, the lowermost horizon of this topsoil still is present. Locally, it is mined as "decomposed granite" for use as fill material and soil conditioners. Old quarry sites are numerous. One can be observed north of Route 78 in Vista. An active quarry operates in Escondido.

The prominent rounded boulders that are so noticeable over much of central San Diego County probably also are a legacy of that period of erosion. Weathering attacked the rocks along cracks, producing residual cores surrounded by disintegrated material. Erosion removed the loose particles, and the residual cores were left as the rounded boulders, which sometimes are precariously perched in dramatic fashion on hillsides. This spheroidal weathering can be observed on Route 78 in Vista and along Interstate 15 between Escondido and Temecula.

Things changed during the Miocene Epoch, about 20 million years ago. The Pacific coast of northwestern Mexico and southern California started to separate. Everything west of the San Andreas fault began to move northwest relative to North America. The breakup was heralded by a renewal of volcanic activity, this time centered farther to the east. Some of those volcanic rocks form isolated volcanic hills such as Cerro de la Calavera in Carlsbad, Morro Hill in Oceanside, Horno summit in Camp Pendleton, and are found as scattered outcrops and plugs in the Jacumba area.

The mountains rose again. The flat plain that had been the west coast tipped up like a giant wedge, forming the Peninsular Ranges from San Gorgonio Pass, east of Los Angeles, to the center of Baja California. Today, as you drive east from Oceanside, you climb the gentle surface of this tipped wedge a few miles east of Julian, and you drop quickly down the steep face of the escarpment to the Salton Trough of the Imperial and Coachella valleys. This rift in the crust is forming as San Diego County splits off from Mexico and moves north. These processes are still going on today, as evidenced by the high seismicity (earthquakes) and geothermal activity of the Imperial Valley.

The sediments derived from these recently emerged mountains were deposited eastward in the newly formed Salton Trough or were carried to the west to be deposited in the Pacific Ocean. By middle Miocene time, a 150-mile-wide strip of land just west of the present southern California coastline was broken up by a series of northwest trending faults into ridges and basins of the Continental Borderland. The sediments of the San Onofre Breccia and Rosarito Beach Formation include distinctive Catalina Island schist and other material that could have come only from the west, indicating that some of the ridges now below the sea were high at that time.

In Late-Pliocene time (approximately 2 to 3 million years ago), the borderland region began to subside and with it the coastal plain of the San Diego region. A fault-bounded basin in the southwestern portion of the county was invaded by the advancing sea and became a broad, shallow bay similar in size and shape to the present Monterey

Bay. This bay, which extended from Pacific Beach to south of the border, supported a diversity of marine life ranging from dolphins and great baleen whales to scallops and lowly mud snails. Through time, the ocean waves cut a broad coastal terrace that today is exposed as the Linda Vista, San Diego, and Otay mesas. This coastal terrace narrows to the north. Long linear remnants stretch along the entire San Diego County coastline. As the sea retreated from this highstand, it deposited on this surface the red, conglomeratic deposits, called the Lindavista Formation that is the curse of local gardeners. Today, this wave-cut platform is covered by streets, homes, and businesses, elevated to a tsunami-safe height.

The last two million years of earth history have been dominated by the advance and retreat of great glaciers. Glaciers never reached San Diego County, but the Ice Ages left their mark indirectly. As ice piled up on the land or melted back into the sea, sea levels rose and fell several hundred feet above and below their present level. One such high stand created 120,000 years ago lasted long enough to create a wave-cut terrace with a deposit of distinctive sediments. These silty and sandy deposits of the Bay Point Formation are easily identified by their drab brown color and intermittent basal gravel.

After the last great ice sheets melted, sea levels rose to their present stage about 6,000 years ago drowning San Diego County rivers. A series of estuaries were created in these drowned river mouths, for example, the lagoons of San Dieguito, San Elijo, Batiquitos, Agua Hedionda, and Buena Vista. The valley of Santa Margarita River contains about 200 to 300 feet of sedimentary fill at the coast.

When sea level was higher than it is now, the waves cut benches into the older deposits, such as the level where Interstate 5 crosses Camp Pendleton. At some locations high stands provided places for shell-rich sediments to be exposed today, such as the west side of Crown Point at Mission Bay.

*Original authorship by the late Dr. Richard P. Phillips (1983). Revised by Tom Demere, Monte Marshall, and Carole Ziegler, with additional contributions by William J. Elliott and Dave Bloom.

STOP 1: Morro Hill
Speaker: Dave Bloom

East side of Morro Hill. Jeep trail ascends to left. Photo: Dave Bloom

The Geologic Atlas of California, Santa Ana Sheet (CGS, 1965) shows Morro Hill as a small stock of Tertiary intrusive (hypabyssal)-rhyolite, with Tertiary volcanic-rhyolite, Tv(r), mapped at three nearby locations to the south and southeast. Larsen (1951) described Morro Hill as a tridymite dacite volcanic plug, and probable source of the flows and pyroclastic rocks of the Indian Hill volcanics of Elliott (1985). The volcanic rocks from this location have been described as Oligocene. However, radiometric dates have not been published for any of the volcanic plugs of western San Diego County. Its age has been interpreted using relative age techniques. As described by Larsen (1951, p. 36), "The rock of the Morro Hill stock is light gray, dense, and finely microgranular in texture. It contains many stout tablets of andesine and much tridymite in part as streaks and in part as needles (probably thin tablets) up to a millimeter long in the mass of the rock. Tridymite makes up about a quarter of the rock." Larsen states that the dacite cuts the Jurassic granitic rocks and some of the Tertiary sedimentary rocks. The Indian Hill volcanics overlie the Eocene Santiago Formation, placing the volcanic rocks younger than middle Eocene.

There are three Tertiary volcanic locations mapped in northwestern San Diego County: Cerro de la Calavera, Morro Hill, and Horno Summit (another

Dacite of Morro Hill with inclusion of granitic rock, approximately 2 inches long. Photo: Dave Bloom.

small "Ti(r)" stock, located about 10 miles northwest of Morro Hill in Camp Pendleton).

Tom Deméré in 2008, implied these Teritary volcanic rocks have a common origin: "...perhaps during the latter part of the Oligocene Epoch (approximately 29 million years ago) renewed regional magmatism caused by shallow plate movements resulted in the extrusion of dacitic volcanic flows and ash from local volcanic vents."

Loose blocks of dacite may be seen in the lower slopes. Hike west up the jeep trail on the south slope of Morro Hill that becomes a foot trail as the slope steepens and turns north, straight up the very steep south slope between prickly pear cactus. Morro Hill rises from an elevation of about 720 feet at the saddle of the road to 942 feet at its summit. Although the horizontal distance is only about 600 feet, over 220 feet of elevation is gained. The trail is loose and a slip here could result in a serious fall that could be even more treacherous on the return. Numerous outcrops near the top of the hill show in-place relations. The massive dacite reveals numerous inclusions of granitic rock.

Morro Hill serves as a boundary for Camp Pendleton, and the northern boundary of the city of Oceanside. A survey benchmark has been placed on its summit. Views to the southwest and north clearly shows the effect the military reservation has on development of the region. While private land has become subdivided and built up to the edge, Camp Pendleton remains largely undeveloped in this area. Following the lower, level road, in the photo shown on page 9 will take you across granitic rocks into which the volcanic neck intruded, and then finally across Morro Hill volcanics to a spectacular

Morro Hill summit survey boundary disk. Photo: Dave Bloom.

westerly panoramic view. So, if you want, you can see all but the survey boundary disk without the strenuous climb.

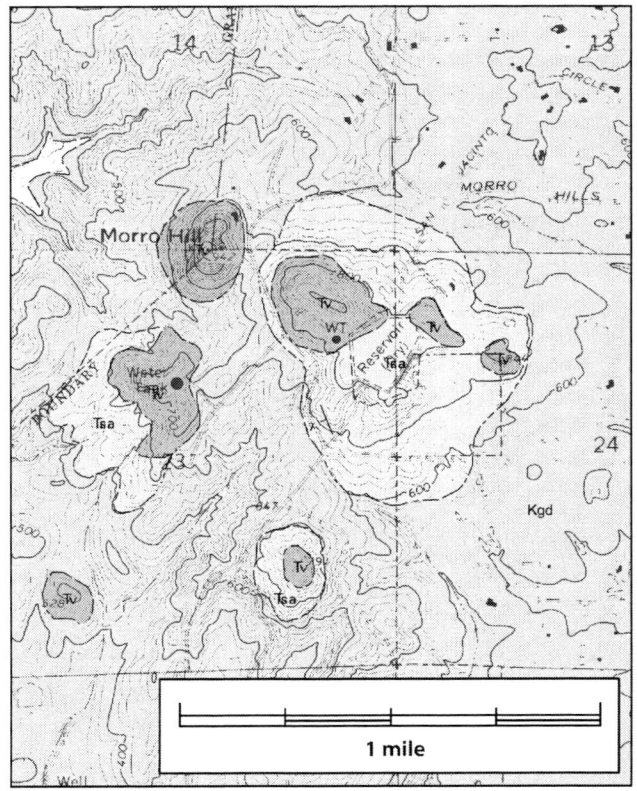

Geology of the Morro Hill Area. Tv=Volcanic rocks undivided (Miocene); flows of dacitic composition. Tsa=Santiago Formation (Eocene); marine sandstone with siltstone interbeds. Kgd=Granodiorite undivided (Cretaceous); mostly hornblende-biotite granodiorite; coarse to medium grained. Reference: Tan (2001).

8.1 Return south on Sleeping Indian Rd to Highway 76 the way we arrived.

11.2 Turn left onto North River Road.

12.0 Weathered Bonsall tonalite in road cut.

13.0 Turn right onto Highway 76 (Mission Road).

16.0 Keep straight ahead through Guajome Lake Road intersection.

21.3 Oceanside Municipal Airport on right. Built in 1963, the airport averages 19 takeoffs or landings daily. The City of Oceanside (2012) approved a 5-year program to modernize facilities.

Benet Road. The white domed building on the hill to the left is the

Rosicrucian Fellowship, "an association of Christian mystics." While Rosicrucians ponder Christian mysteries, we will continue to ponder mysteries of San Onofre, Carlsbad, Encinitas, and the Pacific Rim.

22.9 Turn left onto the Interstate 5 North ramp. The low point in the road before the turn for freeway entrance is about 11 feet above mean sea level, our first place on the trip that is within tsunami inundation hazard area shown on California Emergency Management Agency maps (see page xix).

24.9 Marine Corps Base Camp Pendleton. The second edition of SDAG 1994 Field Trip Guidebook to MCB Camp Pendleton provides information on the history, ecology, and geology of this 195 square mile military facility, established in 1942, that fronts over 10 miles of undeveloped Pacific coastline, the largest in southern California.

25.4 Santa Margarita River. The watershed of this 31 mile long river encompasses 723 square miles in San Diego and Riverside counties, 27% of which is located in San Diego County. The main stem is formed by the confluence of the Temecula and Murrieta creek drainages near Temecula. Santa Margarita River is southern California's only free-flowing river. The lower river and estuary have largely escaped the development typical of other regions of coastal southern California, and are therefore able to support a relative abundance of functional habitats and wildlife. Historically, Santa Margarita was much deeper and wider. Small trading boats actually sailed up the river from the ocean to trade goods behind the Ranch House. The lower river basin shows that during lower sea levels in the late Pleistocene, the river cut a deep valley that has more recently filled with river gravels (Project Clean Water, 2012).

In mid-2013, the old single-track steel truss railroad bridge will be replaced by a modern reinforced-concrete double-tracked bridge currently under construction. According to the SANDAG fact sheet, "The new 755-foot-long bridge will have a 500-foot main bridge spanning the Santa Margarita River and a 255-foot approach trestle spanning the tidal marsh to the south. Torrential floods destroyed the bridge at least twice in the 20th century." (SANDAG, 2012). Maximum flow of the river is reportedly 44,000 cubic feet per second.

28.2 Rest Area. Caltrans rest area at Aliso Creek.

36.1 U.S. Border Patrol San Clemente checkpoint at San Onofre Weigh Station. This checkpoint is located at post mile 67.

39.7 San Onofre Creek. San Onofre was named by the Spanish Padres after the obscure Egyptian Saint Onuphrius.

39.9 Take exit 71, Basilone Road, turn left toward the Pacific Ocean and continue onto Old Pacific Highway. Gunnery Sergeant John Basilone, for whom the road is named, is remembered for his courage and valor in battle at Guadalcanal.* He died in Iwo Jima in 1945.

> *Navy Cross Citation: "Stouthearted and indomitable, Gunnery Sergeant Basilone, by his intrepid initiative, outstanding skill, and valiant spirit of self-sacrifice in the face of the fanatic opposition, contributed materially to the advance of his company during the early critical period of the assault, and his unwavering devotion to duty throughout the bitter conflict was an inspiration to his comrades and reflects the highest credit upon Gunnery Sergeant Basilone and the United States Naval Service." For more information on Basilone, see: www.badassoftheweek.com/basilone.html.

40.3 "San Onofre Beach" artistic sign.

40.7 Turn right at traffic signal. Turn right at the light toward SCE Mesa Facility and proceed to the guard gate. Our next stop is at Nuclear Training Center, located east of Interstate 5. Proceed under the bridge going north on Mesa Road. Mesa Road changes direction after the bridge going south. In 1 mile as you come up the hill, the road splits.

40.8 At the junction is the Nuclear Training Center (EOF). Parking is available on the north, west, and south sides of the training center. We will meet our hosts for this stop.

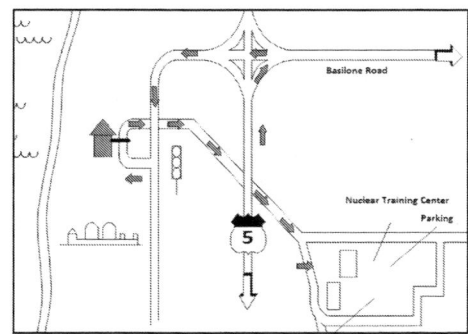

Route to the Nuclear Training Center

STOP 2: San Onofre Nuclear Generating Station
Speakers: Tom Freeman and George Murray

Construction of Unit 1 began in 1964 and first produced electricity January 1, 1968, and operated until November 30, 1992. Units 2 and 3 went online August 8, 1983, and April 1, 1984, respectively, producing 1070 and 1080 megawatts. The plant is operated by Southern California Edison (SCE). San Diego Gas & Electric Company owns 20 percent. The reactors use once-through seawater for cooling so lack the cooling towers characteristic of many other nuclear power plants (U.S. NRC, 2012; SCE 2012).

Extensive nuclear power plant siting studies in the 1960s and 1979s concluded the offshore Newport-Inglewood/Rose Canyon fault (NI/RC) controls seismic hazard at SONGS. SCE is updating its seismic hazard assessment, due March 2014. Studies include geological, geodetic, and geophysical surveys, onshore and offshore. New data on level of activity and history of NI/RC fault will better quantify its total hazard. A better understanding of the existence and characteristics of the "hypothesized Oceanside Blind Thrust" fault is desired. The research will consider specific fault locations, geometries, fault types, slip rates, recurrence intervals, and potential earthquake magnitudes.

Return to the traffic light intersection with Old Coast Highway.

41.9 Turn right onto the main frontage road, Old Pacific Highway.

42.2 Access road on left. Continue straight. Unit 2 and Unit 3 reactors are on the right.

43.2 Enter San Onofre State Beach. Day use entrance fee is $15 per auto.

43.4 Park on the right at Beach Trail 1.

Builders cut into 80,000-year-old marine and nonmarine terrace (dark) and underlying Mio-Pliocene San Mateo Formation (light) (Tan, 1999) to prepare foundation of power plant. Photo: Los Angeles Times, Orange County Edition, *published August 9, 1964.*

STOP 3: Landslides of San Onofre State Beach
Speaker: Mike Hart

The spectacular landslides at San Onofre State Beach extend from near the San Onofre Nuclear Generating Station approximately 4.5 miles south along the beach to a point just north of the rest stop on Interstate 5. The landslides are classified as block glide or translational landslides that are failing along bedding planes in the Miocene Monterey Formation. This formation can be observed in several locations at the base of the cliffs along the beach and consists of dark brown to greenish gray thin bedded siltstone and clayey siltstone.

Overlying the Monterey Formation and occupying the upper more visible parts of the landslides are a thin veneer of marine terrace deposits laid down on the extensive wave-cut platform dated at about 125 ka (marine isotope stage 5e). Overlying the marine sediments are non-marine continental deposits.

Northern reach of the San Onofre landslide complex. Arrows point to landslide headscarps. San Onofre Nuclear Power Plant in upper left.

Oblique aerial photograph, 1954. San Onofre State Beach. U.S. Navy. (Kuhn and Shepard, 1984).

Oblique aerial photograph, 1980. San Onofre State Beach. Landslide measures 700 feet long by 300–320 feet wide, which occurred in the winter of 1978. The small arrows show the direction of movement of landslide debris toward the beach. U. S. Navy. (Kuhn and Shepard, 1984).

One interesting feature of the landslides is that many of them are active and the park service has had to periodically regrade the beach access roads and trails to fill the wide fissures that develop as the slides move toward the beach. According to John Foster who gave a presentation to the South Coast Geological Society on the subject of the landslides, the landslides "toe out" above the beach. While this may be accurate for some of the landslides, observations over several years indicate the formation of bedrock reefs a short distance offshore. These reefs, according to local surfers, have undergone noticeable changes in height over short periods of time. The highly variable bedding attitudes, intense fracturing, and open fissures in the exposed rocky reefs are evidence that at least some landslides extend a few hundred feet offshore. In this case, the triggering mechanism and the reason for periodic reactivation of the landslides is believed to be intermittent unloading of the toe by wave erosion and shifting sand in the littoral zone.

Reef exposed in the littoral zone opposite one of the landslides. The widely variable bedding attitudes, fractures and fissures suggest the landslide toe is offshore. Photo: Mike Hart.

Reset odometer. We return to Interstate Highway 5, and travel south to Scripps Pier.

Odometer (Trip) Mileage

0.0 San Onofre State Beach, Beach Trail 1. Head northwest on Old Pacific Highway.

2.4 Turn right to merge onto Interstate-5 South.

9.9 View Point.

13.4 Aliso Creek rest area.

28.7 Batiquitos Lagoon is located in the city of Carlsbad at the boundary with Leucadia, a community of Encinitas. Batiquitos Lagoon is 600 acres and extends 2.5 miles inland from the coast, where it connects to the Pacific Ocean. The Lagoon collects the flow from a 56 square mile watershed, primarily from San Marcos Creek which extends

14 miles inland, and to a lesser extent Encinitas Creek. Batiquitos Lagoon is technically an estuary, where there is a healthy mixing of fresh water and seawater. The Lagoon is divided into three basins, created by levees of Interstate 5 to the east, and the railroad to the west. The building of California Southern Railroad tracks across the mouth of the estuary in 1881, Pacific Coast Highway adjacent to the Lagoon in 1911, and Santa Fe Railroad tracks across the Lagoon in 1934, tidal action decreased. In 1952 San Marcos Dam reduced the amount of fresh water entering the Lagoon. In 1965 Interstate 5 was constructed across the lagoon. In the 1980s the Port of Los Angeles sought approval of a project to deepen parts of the harbor to 20 feet. To mitigate the loss of coastal habitat, the Port of LA agreed to restore Batiquitos Lagoon. Restoration of the lagoon commenced September 1994 and continued to spring 1997. Tidal flow was restored, dredging enhanced wetlands, and other wildlife habitats were created (Batiquitos Lagoon Foundation, 2012).

30.2 Leucadia Blvd Exit. The west termination of Leucadia Blvd will be our last stop tomorrow at Beacons Beach (Stop 7).

40.9 Road cuts east of the road.

41.5 Slight right to stay on Interstate 5 South at I-805 "Y".

43.0 Take Genesee Avenue exit (exit 29), West.

43.3 Turn right onto Genesee Avenue westbound. Entering the San Diego beach community of La Jolla.

43.7 Science Center Drive. University of California, San Diego on the left. Research institutions and firms, many in the biotechnology industry on the right.

The Regents of the University of California authorized a San Diego campus in 1956. Roger Revelle, then Director of Scripps Institution of Oceanography, led the establishment of the university, selecting the site for the new campus on the mesa next to Scripps and the site of former Camp Matthews. With the support of the citizens and City of San Diego, the university was established in 1960. The first college buildings were completed in 1963. The first class of 181 undergraduates was accepted in 1964. Dr. Revelle recruited a world-class faculty that included Nobel Prize recipients. UCSD swiftly become a leader in scientific research, fueling San Diego's reputation for scholarship and high technology. Dr. Revelle earned an undergraduate degree in geology from Pomona College, and his doctorate in oceanography from UC Berkeley. He coauthored an important paper on the Greenhouse Effect in 1957*, which sparked a whole field of study known as global climate change (University of California, 2012).

*Revelle, R., and H. Suess, 1957, "Carbon dioxide exchange between atmosphere and ocean and the question of an increase of atmospheric CO_2 during the past decades." *Tellus* 9, 18–27.

44.0 Turn left onto North Torrey Pines Road.

44.3 Torrey Pines Scenic Drive. Glider Port.

44.5 Salk Institute.

44.6 Pangea Way.

44.8 Turn right onto La Jolla Shores Drive.

45.8 The building under construction on the left with solar panels on the roof is the new Southwest Fisheries Science Center (SWFSC) Laboratory, replacing the old building across the street.

45.9 View of the peninsula of La Jolla, and La Jolla Bay straight ahead. Southwest Fisheries Building on the right. Much of the facilities of Scripps Institution of Oceanography are being improved in this area as evidenced by the construction projects on this hillside. The original Southwest Fisheries Building, actually four interconnecting buildings, was built in the early 1960s on an unstable 200 foot bluff. Landsliding activity has put the structure at risk. Much of the original building will be demolished, after being replaced with the new building across the street (Geocon, 2010).

46.3 Turn right onto Naga Way, then an immediate left. Scripps Pier is at the bottom of the hill. Parking is limited. If available, park in a metered space. Many parking spaces are restricted to permit holders.

46.4 Scripps Pier, 9321 Discovery Way, San Diego, CA 92037.

STOP 4: Scripps Pier
Speakers: Dr. Neal Driscoll and Dr. Cheryl Peach

The Ellen Browning Scripps Memorial Pier is named for the most significant donor to the institution in its formative years. The original, wooden Scripps Pier was built in 1915–16 and was 1,000 feet long. The new, reinforced-concrete pier, at a length of 1,084 feet, was built in 1988 alongside the original pier, which was then removed. Data about ocean conditions and plankton have been taken from off the Pier continuously since 1916 and provide an unparalleled source of information on coastal Pacific Ocean. In the 1940s the aquarium curator fished from the old pier to catch specimens for display. Small boats can be launched from the far end of the pier for projects in the kelp beds and the Scripps and La Jolla submarine canyons. Seawater is pumped up from the end of the pier, then filtered and stored in holding tanks, providing a supply of fresh seawater to Scripps laboratories and aquariums, including the tanks in the Birch Aquarium at Scripps. The public may also fill containers of sea water for home salt-water aquariums (SIO, 2012).

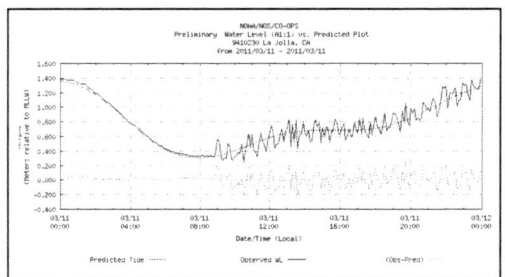

Scripps Pier tide gauge data for March 11, 2011. Tsunami waves appear at 8:54 am PDT. (NOAA, 2012b)

Among the many instruments taking measurements at the pier is a tide gauge. On Friday, March 11, 2011, at 16:54 UTC (8:54 am PDT), the tide gauge jumped 5 cm, and 6 minutes later, another 18 cm, a total of 23 cm higher than predicted. This tsunami was generated 11 hours, 8 minutes earlier, 8870 kilometers across the Pacific Ocean by the Tōhoku earthquake, magnitude, M_w = 9.0. San Diego already received news reports that a large tsunami hit Tohoku, in the Sendai area. The largest runup height was 37.88 meters at Miyako, which flowed 10 km inland. At least 15,703 people were killed, and 4,647 missing (USGS, 2012b). The World Bank estimated losses of $235 billion, the most expensive natural disaster to date. Level 7 meltdowns occurred at three reactors in the Fukushima Daiichi Nuclear Power Plant complex.

Local Tsunami Hazards. Far-field tsunamis, those tsunamis generated from great distances, are of little concern along the San Diego County coastline north to Point Conception. Our offshore rugged topography--the channel islands, basins, banks, and deeps--disrupt tsunami waveforms like baffles so that the waves form, collapse, and reform. Although there can be some damage within harbors, elevation runups are small.

Local-source tsunamis pose a greater damage risk. Scientists at Scripps Institution of Oceanography observed downdropped blocks of sediment along the sides of La Jolla Canyon. Hummocky morphology of slides typical of slumps were observed in high resolution bathymetry. Scripps scientists in collaboration with Steve Ward from University of California, Santa Cruz, made a piston model of slumping. Sandy canyon walls release noncohesive sediment accelerating rapidly to 60 kph into the bottom of the canyon, displacing enough water to generate a damaging tsunami. If the speed of the slump is rapid enough, it could generate a tsunami wave 5 to 8 meters high, striking the shore at 100 kph within 5 minutes.

The biggest uncertainty in this model is how fast a slump accelerates. A rapid acceleration is required, which requires clean noncohesive sand. Validating the model by observing historical tsunami deposits is difficult. Storm deposits and tsunami deposits have many similarities and are difficult to tell them apart. Tsunamis are rare locally. But if you're here at Scripps when you see a tsunami, you don't really need to run fast because the topography is so steep; you just need to calmly grab your beer....

Local tide gauges have recorded tsunami waves since 1854. See Duncan Agnew's updated tsunami history of San Diego (this volume).

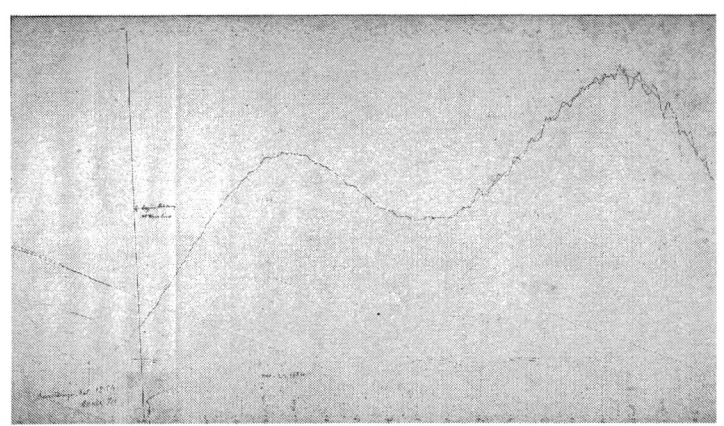

Marigram of the first recorded tsunami in San Diego, December 23, 1854, from an unknown source, shows responses similar to modern records (NOAA, 2012a).

Looking back to shore from the pier, 400 foot high seacliffs rise above the beach. Minch et al. (1998) describes sea cliffs, wave cut benches, marine terraces, and beaches, landforms of marine erosion. As waves pound against shorelines, their impact (6,000 pounds per square inch) erodes uplifted sea cliffs back until a wave-cut bench or platform develops at their base. As erosion continues, the wave-cut bench widens until its width absorbs most of the wave energy and a beach forms along the now low-energy shoreline. The sea cliff diminishes over time due to weathering, erosion, and landslides. In regions of coastal tectonic uplift, like northern San Diego County where uplift rates are estimate to be about 13 to 14 cm per 1,000 years, these wave-cut benches, called marine terraces, are raised above sea level. Generally, the highway will traverse these marine terraces whenever it is near the Pacific coastline.

What's Covering The Rocks? Geology controls biological habitats such as kelp beds and Bristlecone pine forests. One such local example can be seen in the cliffs above La Jolla Bay. Sandstone and siltstone rocks are juxtaposed by the Country Club fault strand of the Rose Canyon fault zone. West of the fault is sandstone which has eroded in typical sandstone fashion with hummocky ledges and pits from meteoric water and wind. This morphology provides habitat for cormorants and other shorebirds. To the east are siltstone cliffs that erode differently. To the west the cliffs are covered by light gray guano. No guano is evident east of the fault on the siltstone cliffs. Dr. Driscoll proposed to call this phenomenon "fault-controlled guano." He especially wanted to share this observation with Prof. Pat Abbott and to thank him for all he has done to make the people of San Diego more geologically aware.

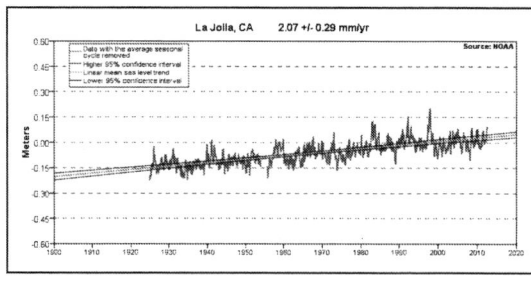

Measurements at Scripps Pier provide a measure of mean sea level trend, which has averaged 2.07 mm per year with a 95% confidence interval of ±0.29 mm per year based on monthly mean sea level data from 1924 to 2006 which is equivalent to a change of 20.7 cm per 100 years. (NOAA, 2012c).

Return to Guajome Park.

Local Damage From Japan Tsunami
Excerpt from Port of San Diego Website
(portofsandiego.org), March 11, 2011

At 4:25 p.m. Friday, a strong tidal surge hit the western tip of Shelter Island, ripping a Harbor Police vessel from its moorings and pushing it against the rocky shoreline. The Harbor Police are currently assessing the damage to the vessel. In addition, debris swept into San Diego Bay during afternoon tidal surges created navigational hazard for boaters, according to Harbor Police Chief John Bolduc.

Tidal surges of between one and two-feet were San Diego's morning introduction to the tsunami triggered by Japan's major earthquake, but stronger surges rocked part of the bay during late afternoon.

Harbor Police reported no damages Friday morning from the tsunami spawned by the 8.9 magnitude earthquake to the San Diego Bay and its shoreline, but Harbor Police Chief John Bolduc warned San Diego boaters in the bay to be on alert for debris.

The tidal surges in Shelter Island were recorded between 10:15 a.m. and 11:15 a.m. Following the surge events, the Harbor Police conducted a sweep of all San Diego Bay anchorages and marinas and found no damages.

A number of precautionary measures were taken as a result of the tsunami advisory.

The Carnival Spirit cruise ship was due to arrive in San Diego at about 8 a.m. Friday, but was delayed for nearly three hours. Morning fog was responsible for some of the delay, but later, the U.S. Coast Guard advised the vessel to remain off the coast. After the Coast Guard gave clearance, the vessel was escorted by Harbor Police vessels into San Diego's B Street Cruise Terminal, arriving at about 10:45 a.m. The vessel carried about 2,500 passengers who were traveling from Cabo San Lucas to San Diego.

While docked in San Diego, extra ties were used to secure the vessel and its anchor was dropped as hedges against a tsunami surge.

Port officials directed a fuel barge that was docked at 7:30 a.m. Friday morning at the Port District's 10th Avenue Marine Terminal to reposition itself in the bay's channel until further notice. When the tsunami threat passed it was allowed to return to the terminal later in the morning.

Several Navy vessels were re-deployed to a site in the South Bay away from their piers.

Tsunami damage in San Diego Bay from the February 27, 2010, Chilean earthquake (M_w 8.8), the fifth largest earthquake to strike during the age of seismological instrumentation.

Chilean Quake Tsunami Caused Damages in San Diego Harbor
John Campbell
Campbell Pacific, 3960 W. Point Loma Blvd, Ste. H 347, San Diego, CA 92110
johncinsd@cox.net

There was a bit of everything in San Diego County Saturday, February 27, 2010: rain, hail, lightning, snow to 3,500 feet, sun, wind gusts to 20 knots, and a tsunami of what looked to me like 2 to 3+ feet, but a gentle one, not a wall of water. I watched over 15 sailboats start a race at noon off of Harbor Island, heading into a 20 knot westerly wind and a very dark sky looming over Point Loma. Other than that, the harbor was quiet; no other noticeable traffic.

I got a call about 1:30 pm from Captain Fayette "Bubba" Severence of SeaTow saying a pier at Driscoll's Wharf in America's Cup Harbor was breaking up. I jumped in my car to check it out and offer my assistance.

Capt. Bubba was already on scene with the SeaTow 35-foot Eastern when I arrived at Driscoll's Wharf in pouring and driving rain. He had a line onto the 110 foot research vessel *Merlin*, a 1981 Swiftship, which had been end tied to the dock. He waited until the current had ebbed and was starting backout and then he was able to get the ship out into the bay and set it at anchor. The remainder of the crew arrived and he picked them up and shot them out to the ship. Later, Capt. Bubba told me the captain of the Merlin requested assistance because he believed his props were fouled with lines and debris. This was later confirmed when the engines were engaged.

The *Merlin* had received damage to its aluminum hull on the port side. The ship was relocated to the Driscoll yard on Shelter Island Drive. If the *Merlin* had broken off and was not underway, it could have wreaked untold millions of dollars in damage to the piers and moored boats, many of which are live aboards.

SeaTow Capt. Bubba then rescued a sailboat aground just outside of America's Cup Harbor on a sandbar. After that he assisted his fellow SeaTow rescue boat take in-tow a 70-foot ketch that had dragged its mooring ball anchor block 20+ feet and was in danger of damaging other boats.

Capt. Bubba called me, about 3 pm and asked me to try to find Larry Baumann, owner and general manager of the Bali Hai Restaurant, because it looked like his new docks were damaged and partially missing. I went over to the Bali Hai and unable to find Larry. I went down to check the docks and found his son, Andy Baumann. There was a 10-foot section missing from the dock, just seaward of the gangway. The surge must have been very strong and pushed the dock underwater; the force of the water breaking off the section. While observing the dock, I witnessed two surges I guessed they were 10 to 12 knots both incoming and outgoing.

As I think back on the day, I'm stunned at the lack of awareness and action by the boating community, the Coast Guard, and the Harbor Master for not making sure able boats left the harbor when it was determined a tsunami was eminent as it approached San Diego.

Is it not common knowledge the effects and destruction a tsunami can cause by eddies in a harbor? After 28 years in Hawaii as a yachtsman, I witnessed what is done when there is a tsunami alert. As many watched on television, the harbors in Hawaii, including the Ala Wai Yacht Harbor and Pearl Harbor, had all able boats and ships depart for the safety of the open ocean.

The U.S. Coast Guard in San Diego on Saturday advised boaters to remain in the harbor due to small craft warnings. Coast Guard spokeswoman Jetta Disco said, "The San Diego area experienced tidal surges of 4 to 5 feet for about 90 minutes. Some boats in Mission Bay, San Diego Bay and Oceanside Harbor broke free from their moorings, but were later re-secured."

History of Guajome Regional Park Land and People
by Eleanora I. Robbins
San Diego State University
norrierobbins@cox.net

5,000–7,000 years: Land of the Luiseño people, who are Shoshonean speakers, part of Uto-Aztecan stock; they call themselves Payoomkawichum ("people of the west"). They ate and used many of the plants that are still here: acorns, chia sage, cacti, yucca, berries, and bulbs in the lily family. Meat protein came from rabbits, ground squirrels, woodrats, mice, deer, lizards, and ducks. The bigger animals who also lived here were deer, wildcat, mountain lions, and grizzly bears; excavated bison bones reveal the Pleistocene presence of buffalo. For music, the people created bullroarers, flutes, cane whistles, and tortoise shell rattles. Women used the wetland vegetation to make baskets, whereas men made nets from Indian hemp and milkweed. When Europeans arrived, they found the people living along San Luis Rey River. Five villages have even been identified in the park. The name of the park either comes from the village Wajoma or from "wakhavumi," which means frog pond, which is still here.

53 years: Spanish territory (1769–1822). Mission San Luis Rey de Francia was built in 1798. The Franciscan missionaries and their diseases disrupted the lives of the Luiseño people, who were brought into the mission system to work the crops, and herd the cattle and sheep.

24 years: Mexican territory (1822–1846). The Luiseño people next became indentured to Mexican feudal barons. Governor Pio Pico created the first land grant in 1836 and gave Rancho Guajome to two Luiseño brothers, Adrés and José Manual. Shortly afterward, they sold it to Abel Stearns. The Mission was abandoned in 1845.

158 years: United States (1848–present). After California became a state, the Land Claims Commission decreed the sale of Rancho Guajome to be illegal. Cave Couts, who played prominently in the early political history of the U.S. here, arrived in 1849; he was a West Point graduate, sent to reinforce troops in California. He befriended Don Juan Bandini, and met and married Bandini's daughter in 1851.

Stearns gave Rancho Guajome as a wedding present to his sister-in-law, Ysidora Bandini. Couts built the Guajome Ranch House in 1852–53. On this land he planted orange, apple, peach, almond, pear, quince, fig, lemon, pomegranate, black walnut, apricot, mulberry, plum, olive, persimmon, cherimoya, mango, and avocado trees. In 1858, the Luiseño brothers theoretically got the title back to the land, but not in actuality.

The years 1862–63 were one of the most damaging times to the land, the people, and the animals. The smallpox epidemic of 1862 killed many of the Indian people and a winter deluge destroyed a quarter of California's taxable wealth. The following spring, grasses next grew abundantly, cattle proliferated, driving prices down. A two year drought followed. Couts, whose wealth fluctuated with these events, died in 1874.

The Ulysses S. Grant administration made a decision in the late 1860s to move Indian people away from cities and into reservations. Most Luiseño people were moved between 1875 and 1892. In 1880, Helen Hunt Jackson came to San Diego to learn more about treatment of the Indians; she visited the Guajome Ranch House and was appalled by what she saw. She published the results of her research in *A Century of Dishonor* in 1881. This was followed by *Ramona* in 1884. Some think the Guajome Ranch House was the model for the rancho in *Ramona*.

In 1970, the Guajome Ranch House got National Historic Landmark designation. San Diego County acquired the property in 1973 through a condemnation proceeding and created Guajome Regional Park. The park, which is four miles from the mission, sits at the boundary between two cities, Oceanside which was incorporated in 1888, and Vista which was incorporated in 1963. Like all the people living around the park, Luiseño people play an active role today in urban, reservation, mission, educational, political, and musical affairs.

In 2011, a monument was created in the form of a Luiseño basket, which is now displayed at the center of the park.

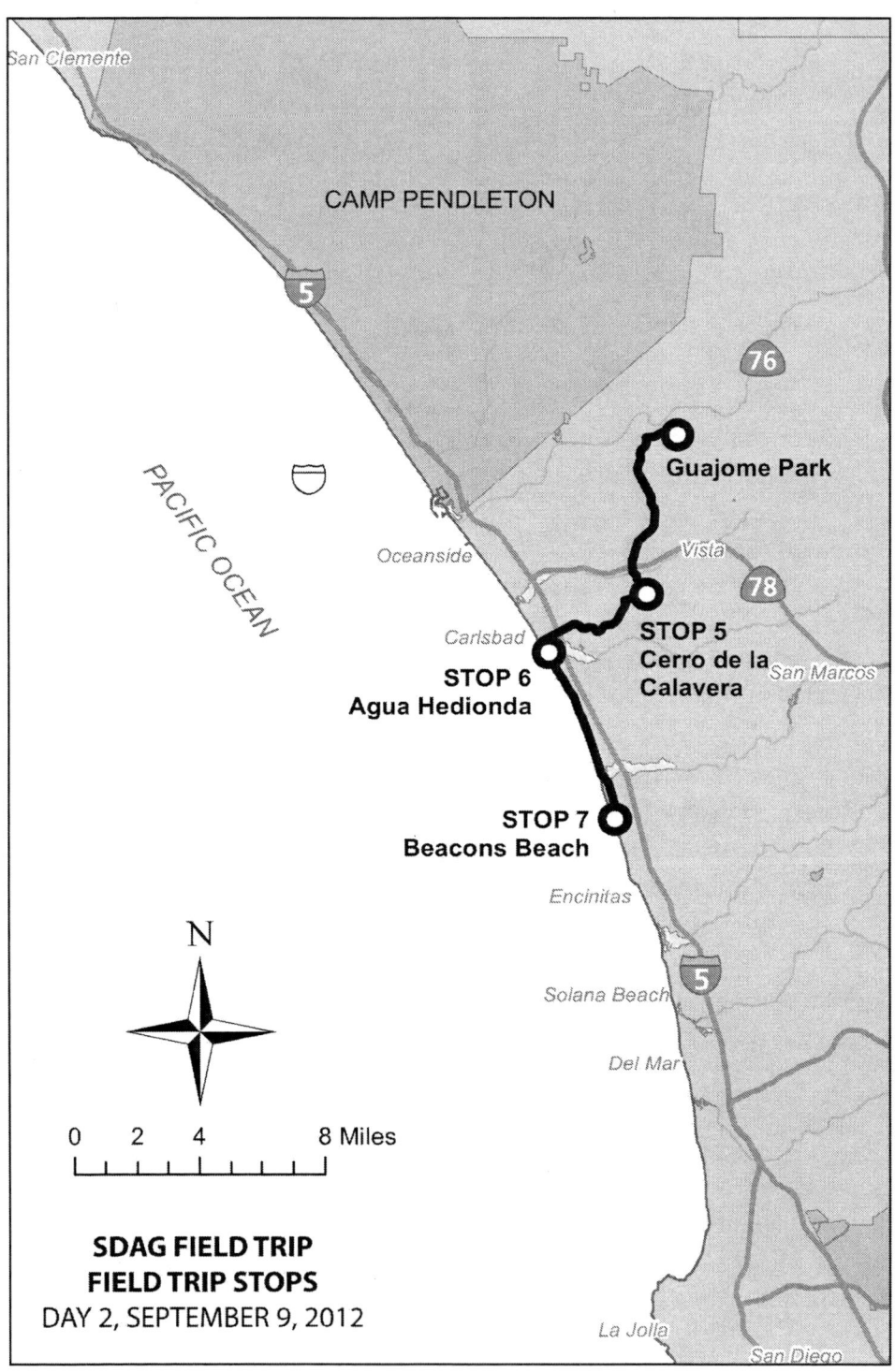

DAY 2: SUNDAY, SEPTEMBER 9, 2012

Begin at Guajome Regional Park. We will travel by private car today. Be sure to carpool. Limited parking will be available.

Leave our campground at Guajome Regional Park. Reset odometer at the park entrance on Guajome Lakes Road.

Odometer (Trip) Mileage

0.0 Exit Guajome Park. Turn left on Guajome Lakes Road.

0.1 Turn left onto Highway 76 (San Luis Rey Mission Expressway). Turn left onto College Blvd.

5.5 Entrance to Mira Costa Community College.

6.1 Marron Road. Lake Road. On the right is a large hillslope cut for the Walmart store and parking lot. The gray rock is Green Valley tonalite. According to Larsen (1951, p. 20),

> "This tonalite is not very different chemically and mineralogically from the Bonsall tonalite, but it lacks the abundant streaked inclusions of that tonalite. The Green Valley tonalite underlies an area of about 68 square miles in the San Luis Rey quadrangle. It intrudes the Triassic sediments, the Jurassic volcanics, and the San Marcos gabbro, and is intruded by the Woodson Mountain granodiorite and several other granodiorites.
>
> "The rock is gray, medium-grained, and rather uniform in character. The minerals, especially the dark minerals, do not show a sharp granularity in the hand specimens. The plagioclase crystals are gray from inclusions. Under the microscope, the average rock shows a variable grain size, ranging from half a millimeter to 3 millimeters."

6.6 North Tamarack Ave. Go straight here. Turn left at the **next** Tamarack Ave.

6.7 Turn left onto Tamarack Ave. The sign says the residential neighborhood is "Calavera Hills." We are going to Cerro de la Calavera, which is visible about ½ mile ahead and to the right.

7.0 Go approximately 0.3 mile and park on the right, or park on Strata Drive if no parking spaces are available.

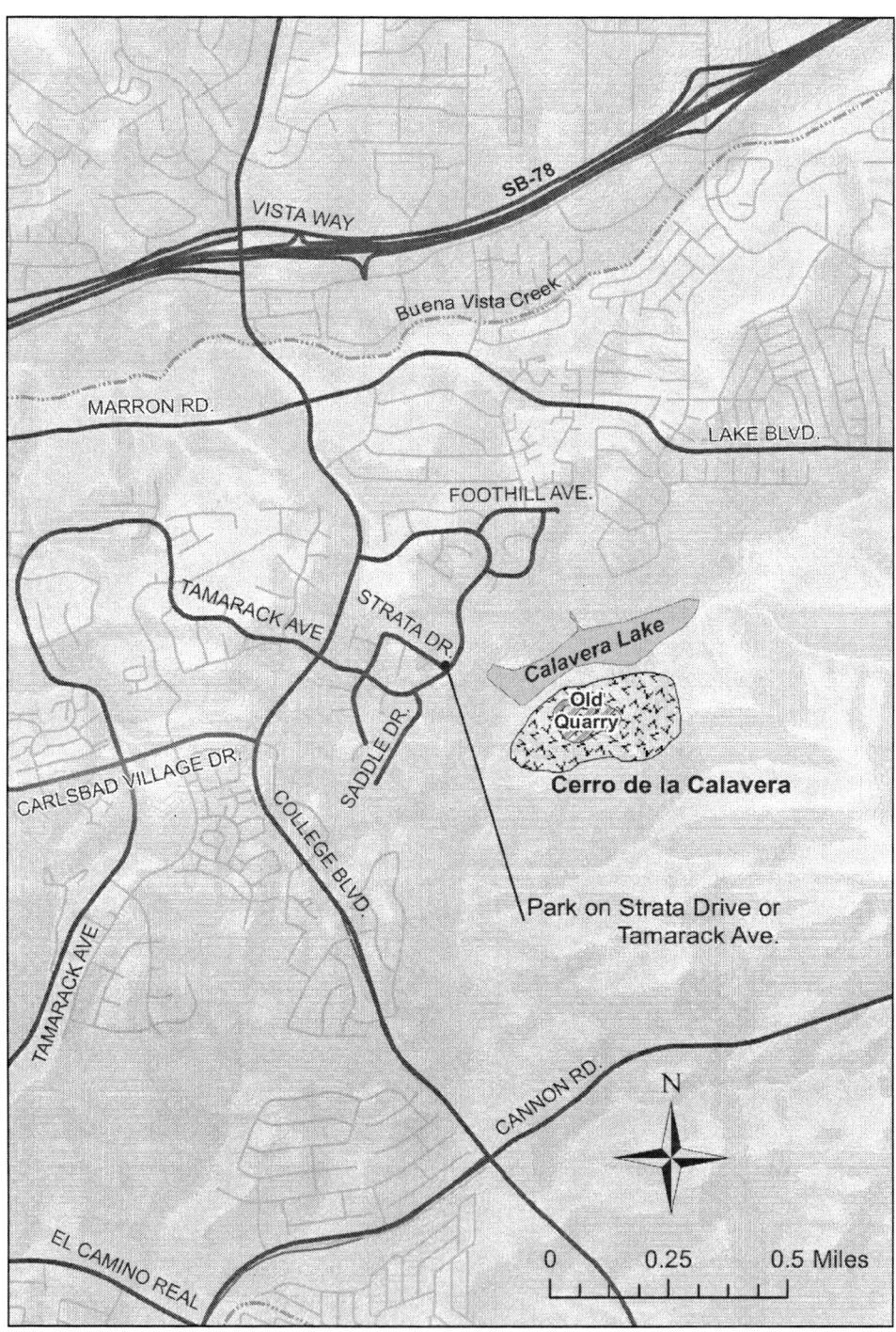

Park at the intersection of Tamarack Avenue (not Tamarack Ave North) and Strata Drive.

STOP 5: Cerro de la Calavera
Stop leader: Phil Farquharson

Cerro de la Calavera is the best exposed and most accessible of the four Tertiary volcanic exposures in western San Diego County.

John Turbeville, professor at Mira Costa College, has written a paper on the rocks in the Cerro de la Calavera (Skull Hill) area. We will follow the route of Turbeville (2008), which takes us on a four stop 1.5-mile round-trip hike across the dam that creates Calavera Lake, to the quarry site in the volcanic rocks, and to two exposures on this side of the hill. There is about 300 feet of elevation change, and part of the hike is moderate to difficult.

Turbeville states the volcanic rocks are generally intermediate in composition and form distinct flows along the flanks of the extinct volcano, and are thought to have formed in the Miocene. This is a time when our coastline was transitioning from a convergent plate boundary (subduction) to a transform plate motion. This transform motion eventually formed the San Andreas fault. The volcanic processes here are thought to be similar in age and genesis to other volcanic events occurring up and down our coastline during the Miocene.

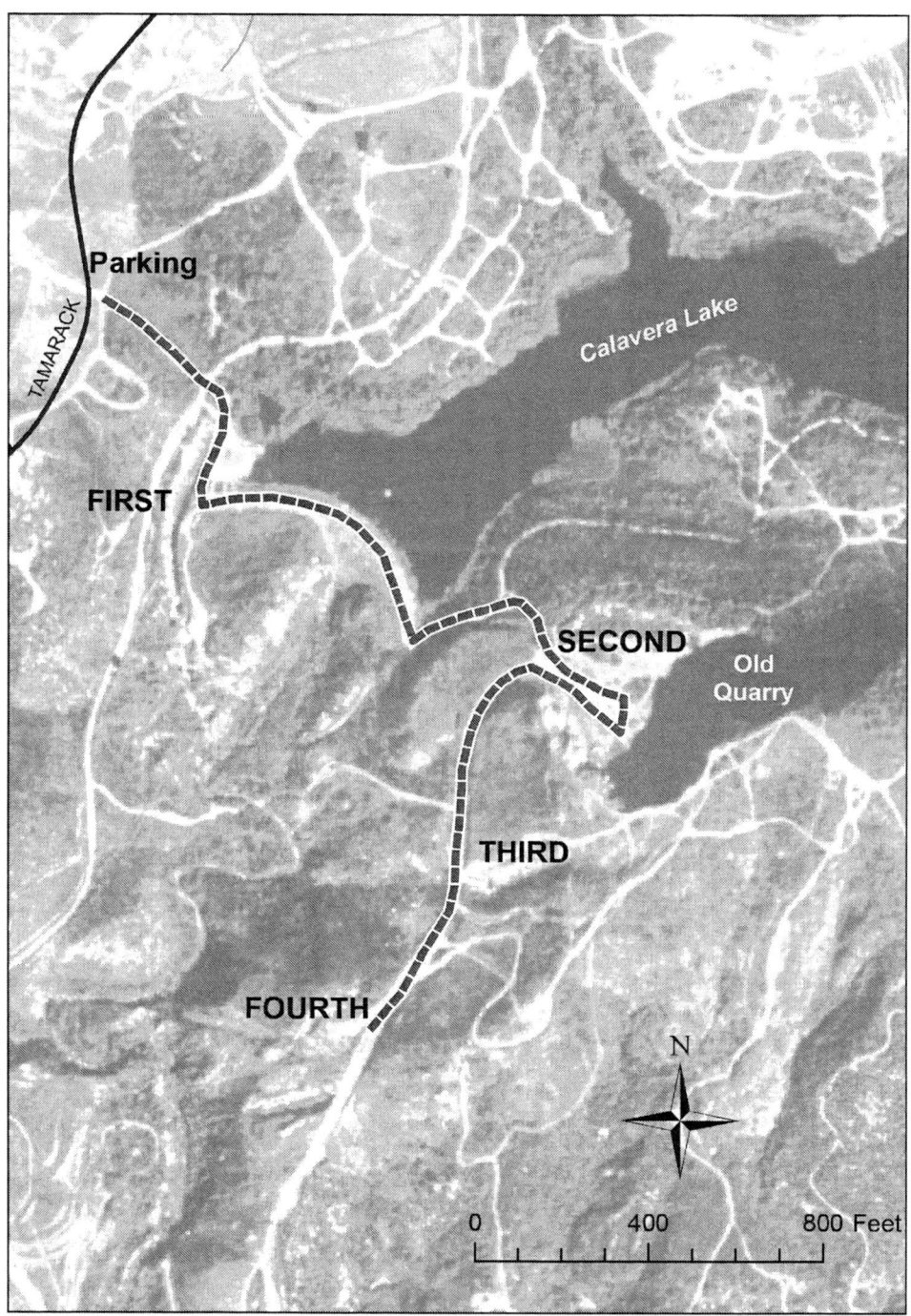

The four stop hike described by Turbeville (2008) takes one down to the lake up a steep trail to the quarry, along a dirt road, and back to the parking area.

From Turbeville,

"The volcanic rocks cut the Eocene Santiago Formation (Tan 1996). This makes them younger than Eocene with a likely window of activity between late Oligocene and Miocene. This timing correlates well with other similar regional volcanic activity that occurred along our coastline (to the north and south of us) during the same timeframe. One example of this is the Conejo Volcanics found on the western side of the Santa Monica Mountains just north of Los Angeles (Weigand 2002). In addition, there is a great deal of volcanic ash found locally in sedimentary rocks of this age. It has not been determined which of the volcanic plugs (or all) located in this area produced the ash (Abbott 1999)."

Contact between dacite rocks of volcanic plug (Tda) and Eocene Santiago Formation (Tsa). Distinct baked zone at contact. Between Turbeville's second and third stops. Photo: Dave Bloom.

The tectonic framework of this area is transition from subduction to transform. Turbeville continues,

"Extensive subduction occurred on the western edge of North America from early Mesozoic to early Miocene (approximately 200 to 19 million years ago). Oceanic crust equaling the size of the entire Pacific Basin was subducted underneath the North American plate during this time.

A transition from subduction (and plate convergence) to primarily a transform boundary occurred during the Miocene (from approximately 19 to 5 mya). This transform boundary eventually became the San Andreas fault (Atwater 1998). During the transition a spreading center was obliquely subducted along the convergent boundary. As the warm, buoyant spreading center was subducted the overlying continental crust was uplifted, fractured, and parts were significantly rotated. Some of rotational events were fairly large; one of them is believed to have formed the Transverse Ranges to the north of Los Angeles (Fritsche, 2001). This series of events left zones of weakness in the continental crust where magma could ascend to the surface to become lava. The volcanic plugs found in northern San Diego County are thought to be a product of this tectonic transition from subduction to lateral motion."

Turbeville describes the three main rock units that are observed in this area, ranging from Cretaceous to Miocene in age:

Green Valley Tonalite (Cretaceous)(Larson 1948): The first rock unit we will see is part of the Peninsular Ranges. It is an intrusive igneous unit that cooled slowly at depth and subsequently uplifted, eroded and exposed. It is mapped as a tonalite, which is similar to a granodiorite.

Santiago Formation (Eocene)(Tan 1986, 1987) In this area, the Santiago Formation is very soft and for this reason there are only a few outcrops in the area. The local extent of the formation is primarily based upon the light colored material it generates from weathering. The Santiago Formation was deposited in a beach setting. The sandstone is fine to medium grained, poorly cemented, poorly bedded, and is interbedded with siltstone and claystone that is locally prone to significant slumping. The Santiago Formation is not to be confused with the Santiago Peak Volcanics that are much older.

This formation was renamed from the Scripps Formation locally in the mid-1980s (Tan 1986, 1987). It is found regionally from Orange County to northern San Diego County where it begins to interfinger to the south of Leucadia with the Torrey Sandstone.

Volcanic Plug (Miocene): This rock unit is "mouse gray" in color with an aphanitic texture, which means that you cannot see the grains with the naked eye. The rocks from the volcanic neck are dacite, an intermediate composition between rhyolite and andesite.

The geologic map of the Oceanside 30' × 60' Quadrangle, California (Scale 1:100,000) shows Cerro de la Calavera mapped as Tda "Dacite stock (Miocene)" surrounded by Santiago Formation (Tsa) to the north, east, and south, and tonalite, undivided (mid-Cretaceous) (Kt) to the west. The interactive version of this map shows the geologic map overlain on a Google map, so modern roads and other culture are visible.

Preserve Calavera (2012) says, "The 513-foot Mount Calavera is not really a mountain at all but rather a 22 million-year-old volcanic plug. Mount Calavera is one of only three volcanic plugs in Southern California. In the early 1900s, the ancient plug was mined for gravel. The mining was accomplished by stripping away its west face and continued into the 1930s." A report on the building of the Calavera Dam implies the quarry was used at least until 1941 when the dam was built (Carlsbad Historical Society, 2006).

There is no reference a 22-million year age for the volcanic rocks of Calavera. *North County Times* reports Dave Kimbrough and Tom Demére planned to obtain radiometric dates on the rocks and several other volcanic rocks in the San Diego region. Tom Demére is quoted as making an initial estimate of 28 million years ago for the age of the rocks. John Turbeville is said to be planning his own age-dating of the rocks. Apparently, results from dating the rocks have not been published.

Tom Demére (2008) implies the volcanic rocks of Cerro de la Calavera:

> "...perhaps during the latter part of the Oligocene Epoch (approximately 29 million years ago) renewed regional magmatism caused by shallow plate movements resulted in the extrusion of dacitic volcanic flows and ash from local volcanic vents."

It would be interesting to compare and contrast the volcanic rocks of Cerro de la Calavera, Morro Hill, and Horno Summit.

Some of the use trails are fenced, with signs telling visitors to keep out, sensitive habitat restoration in progress.

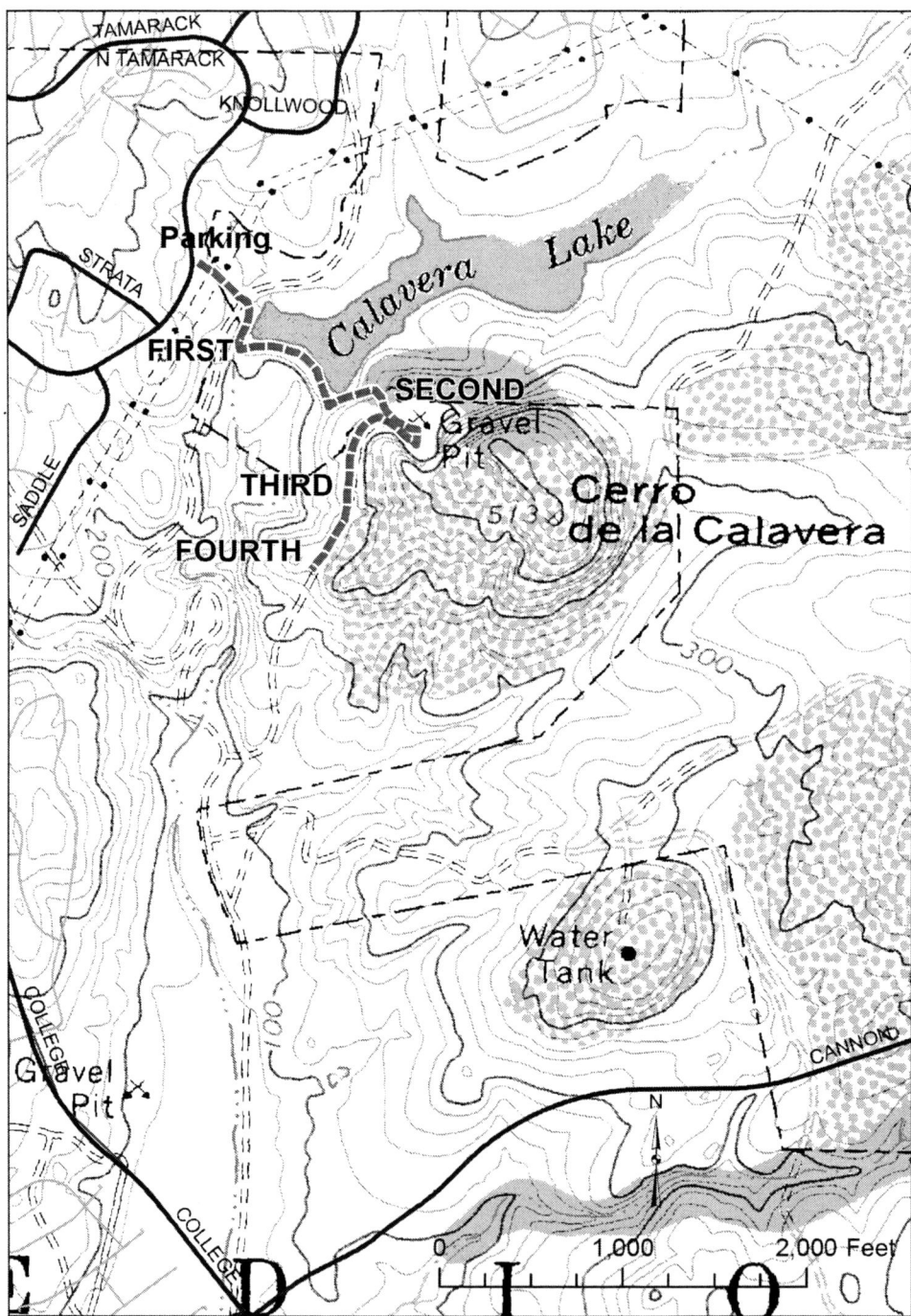

Topographic map of Cerro del la Calavera showing four-stop hike relative to Tamarack Avenue, Calavera Lake, the old quarry, and vicinity.

Concrete paved spillway obscures fresh tonalite of Turbeville's first stop. Weathered tonalite visible above the concrete. Photo: Dave Bloom.

Turbeville's Stop 1 is a drainage ditch showing unweathered tonalite at the bottom and highly weathered tonalite higher up. The ditch is totally paved with a concrete spillway, so the tonalite will need to be observed elsewhere.

There are many other places to observe the Cretaceous Green Valley tonalite. It is unclear why the ditch was lined. If the bottom of the ditch was fresh tonalite, it seems no great benefit was made in paving it. The Green Valley tonalite was dated at 120 m.y. using lead-alpha (Bushee and others 1963), and a later lead-alpha date of 90 ± 10 m.y. from zircon in San Pasqual 7.5-minute quadrangle (Marvin and Dobson, 1979). According to the National Geologic Map Database (USGS, 2012a), the first use of the name Green Valley tonalite was by Miller (1937), and described by Larsen (1948) as a gray medium-grained tonalite, uniform in character. Resembles Bonsall tonalite but lacks abundant streaked inclusions. Intruded by the Woodson Mountain granodiorite. Everhart (1951) described it as older than Bonsall tonalite and younger than Cuyamaca gabbro.

Cross the dam that forms Calavera Lake. Impounded water stored here is prevented from flowing about 3 miles downstream to Agua Hedionda Lagoon. Carlsbad Historical Society (2006) provides historical photographs showing the construction in 1940–41. The text says this location was chosen for the dam, "because it contained white silica sand, which meant a minimum of seepage and because of its proximity to rock and sand from a nearby quarry." This area is popular for fishing, hiking, and walking dogs.

Hike up the trail to the quarry, past mudstone mapped as Eocene Santiago Formation. Ascend to a flat at the base of the quarry, Turbeville's second stop. Columnar jointing is present in the volcanic rocks toward the top of the quarry. Talus covers most of the lower part of the quarry. The volcanic rock is massive and aphanitic. Few black euhedral crystals (hornblende?) were observed. The trail up to in-place rocks is steep and footing is uncertain on the loose talus. Hiking the talus is risky.

Columnar jointing of dacite plug of Cerro de la Calavera at the second stop of Turbeville. Photo: Dave Bloom

Between Turbeville's second and third stops, the contact between volcanic rocks and Santiago is well expressed. The "adit" or cave at Turbeville's third stop is curious. The "lava flow" of Turbeville looks like recent landslide breccia to this observer. It looks like a landslide over the tonalite. The "rocks" to the right of the opening are obviously concrete, with rebar.

Turbeville's fourth stop may be difficult to find, but inclusions of granitic rock are present in the volcanic rocks. In the road cut, tonalite core stone and highly weathered "sap rock" were nicely exposed. Even in the road itself, granular weathered tonalite showed concentric bands of iron oxidation (dark brown and purple). Feldspars appeared to be largely altered to clay.

Highly weathered Green Valley tonalite expressed by concentric bands of iron oxide, southwest slope of Cerro de la Calavera, fourth stop on hike. Width of view is approximately 20 cm.

7.0 Return to your car, turn back southwest on Tamarack Ave.

7.3 Turn left onto College Blvd.

7.5 Take the 1st right onto Carlsbad Village Dr.

7.8 Turn left onto Tamarack Ave.

11.3 Turn left onto Carlsbad Blvd. This road is County of San Diego Route S21, known also as historic U.S. Highway 101. Commissioned in 1926 as one of the original U.S. highways, this portion of Highway 101 was decommissioned in 1964 when it was replaced by Interstate 5. The designation "Historical U.S. 101" with distinctive brown shield signs reminiscent of the original U.S. 101 signs recognizes the historic importance of this highway.

11.9 Park along the highway at the beach, with Agua Hedionda on the left.

STOP 6: Agua Hedionda Lagoon
Speaker: Phil Rosenberg

Agua Hedionda Lagoon ("smelly water lagoon") has been extensively modified at its connection with the ocean by the coastal highway, railroad, and Interstate 5, the Encina power plant, aquaculture farm, and proposed desalination plant.

The Encina power plant was constructed in 1954 by San Diego Gas & Electric Company as an oil fueled thermal plant. The power generating facility, now owned and operated by NRG Energy, Inc., now uses natural gas to produce up to 965-megawatts of electricity from 5 units (NRG, 2012). The facility uses once-through cooling from an intake channel at the north side of the lagoon, and discharge through a channel at the south side, separated by ½ mile bay-mouth bar. Constant flow of high quality sea water through this basin of the lagoon has made conditions well suited for aquaculture (Le Page, 2007).

On May 31, 2012, the California Energy Commission approved the construction of a proposed 558-megawatt combined-cycle natural gas power plant next to the existing power plant (California Energy Commission, 2012).

Agua Hedionda is one of several important lagoons that provide valuable intertidal estuarine flats. Although Agua Hedionda is one that has been greatly altered by industrial development, in general San Diego County's tidal lagoons are vegetated by various forms of algae, and are fringed with a growth of intertidal Pickleweed and other halophytes. Minch et al. (1998) describe tidal marsh environments in general for the region:

> "Tidal flats are the home or favored resting place for many living organisms adapted to life in this extremely harsh saline environment. Organisms that inhabit tidal flats face severe changes in daily environments. They are alternately immersed by high tides or exposed by low tides. Part of the tidal flat will always be underwater in sloughs and channels, and part will be above the high water mark. Because of these changing conditions, variations in soil salinity are to be expected.
>
> "In the flattest lowest portion of the mud flats the salt concentrations may exceed 6% which prevents the growth of all plants except marine algae. Near the shallows and edges of the flats, salinity both decreases and fluctuates with fresh water runoff and rain. Temperature ranges are broad which necessitates adaptations for surviving fluctuations of as much as 50° F in a single day's cycle."

Typical dune and strand line halophytes include sand verbena, sand bur, beach evening primrose, beach fig, door brush, and ice plant. Typical salt-marsh halophytes are yerba mansa, salt grass, alkali heath, heliotrope, spiny rush, sea lavender, salt cedar, glasswort, pickleweed, sea purslane, and sea blite.

Up slope from the influence of sea water, less salt tolerant flora of the Coastal Sage Scrub community dominate the undeveloped hills and mesas.

Estuarine lagoons mix nutrients from freshwater streams and seawater tidal flows, providing productive habitats for animals to prosper. Tidal flats are especially favorable to birds, such as pintails, mallards, coots, gulls, terns, curlews, willets, dowitchers, yellowlegs, and whimbrels that feed on molluscs, worms, and shrimp that thrive in these intertidal environments (Minch et al., 1998, p. 22–23).

Salt Marsh Plants

1. *Anemopsis californica* — Yerba Mansa
2. *Distichlis spicata* — Saltgrass
3. *Frankenia grandifolia* — Alkali Heath
4. *Heliotropium curassavicum* — Salt Heliotrope
5. *Juncus acutus* — Spiny Rush
6. *Limonium californicum* — Western Marsh Rosemary
7. *Monanthochloe littoralis* — Shoregrass
8. *Salicornia subterminalis* — Pickleweed
9. *Salicornia virginica* — Glasswort
10. *Sesuvium verrucosum* — Sea Purslane
11. *Suaeda esterea* — Estuary Sea Blite

11.9 Resume south on Carlsbad Blvd

16.4 Leave city of Carlsbad. Enter city of Encinitas, community of Leucadia. Continue on North Coast Highway 101.

17.6 Turn right onto West Leucadia Blvd.

17.8 Turn right on Neptune Ave and turn left into the Beacons Beach parking lot.

STOP 7: Beacons Beach
Speaker: Dave Schug

Beacons Beach Landslide, Encinitas

The coastal bluff at Beacons Beach has experienced historic instability associated with a large landslide. Landslide movement had damaged previous beach access stairways during winter storms of 1982–83. Since that time, access to the beach is via a switchback trail leading down from the parking lot at the west side of Neptune Avenue. The coastal landslide encompasses virtually all of the coastal bluff below the parking lot. The beach access trail has been repaired and rerouted many times to avoid steep, hazardous slopes.

The coastal bluffs along Encinitas are comprised of Eocene sedimentary formations that are relatively resistant and form near-vertical seacliffs. Pleistocene terrace deposits making up the upper bluffs form more moderate slopes. When oversteepened, the terrace sands slough back to flatter slope inclinations.

Initial landslide movement at Beacons was strongly influenced by wave erosion that had undercut weak claystones along the toe of the bluff. Bedding plane shears in the area occur at a low elevation that was frequently exposed to wave erosion. Lowered to non-existent beach sand levels had resulted in wave notching at the seacliff toe, contributing to a weakened condition. A back-rotated slide occurred mostly along the weak bedding planes. Since that time, additional slumping, and subaerial erosion has taken place within the sandy terrace deposits. The current bluff edge has eroded landward up to the edge of the parking lot.

Beach sand levels and other factors, including drainage and seismic shaking influence the general stability of the coastal bluff at Beacons. Beach replenishment projects beginning with the Regional Beach Sand Project (RBSP I) in 2001 increased the width of the beach at Encinitas. The RBSP II beach replenishment project, underway as of Fall 2012, should further reduce erosion potential along the Encinitas coastline.

Two oblique aerial photograph views of landslide at Beacons Beach, Encinitas. February 2002. Photos: URS.

1957 Oblique Aerial Photograph. Beacons Beach. Source: San Diego Historical Society, TICOR Collection.

2002 Oblique Aerial Photograph at nearly the same scale and orientation to compare with the 1957 photograph. Beacons Beach. February 2002. Source: URS.

Highly fractured blocks of Ardath Shale are back-rotated at the slide toe. Note cobble berm along the back beach. Exposed Landslide Toe. Beacons Beach, circa 1989. Photo: URS.

Probable basal landslide surface exposed above the cobble berm. Exposed Landslide Toe. Beacons Beach, circa 1989. Photo: URS.

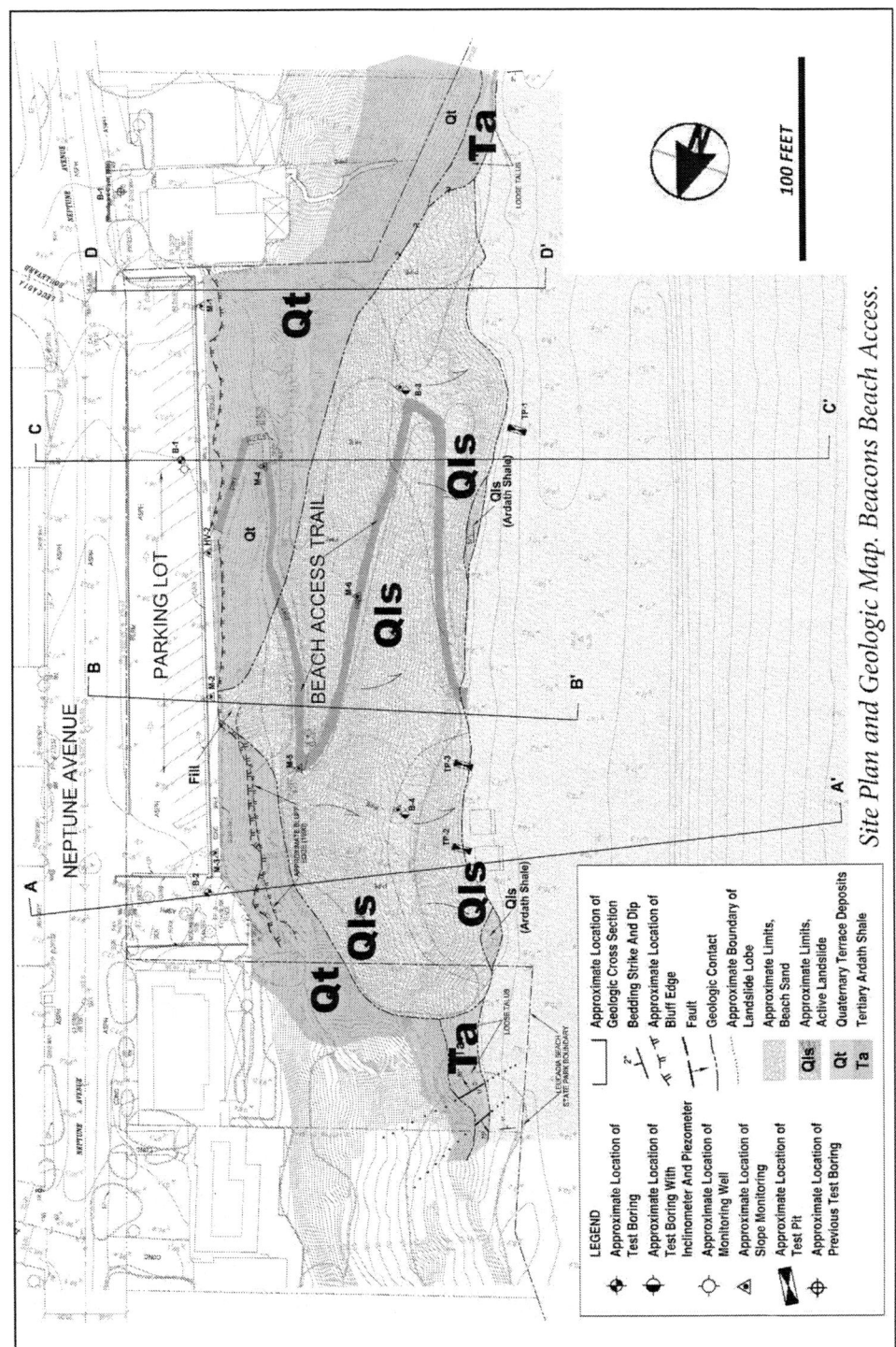

Site Plan and Geologic Map. Beacons Beach Access.

WAITING FOR TSUNAMI

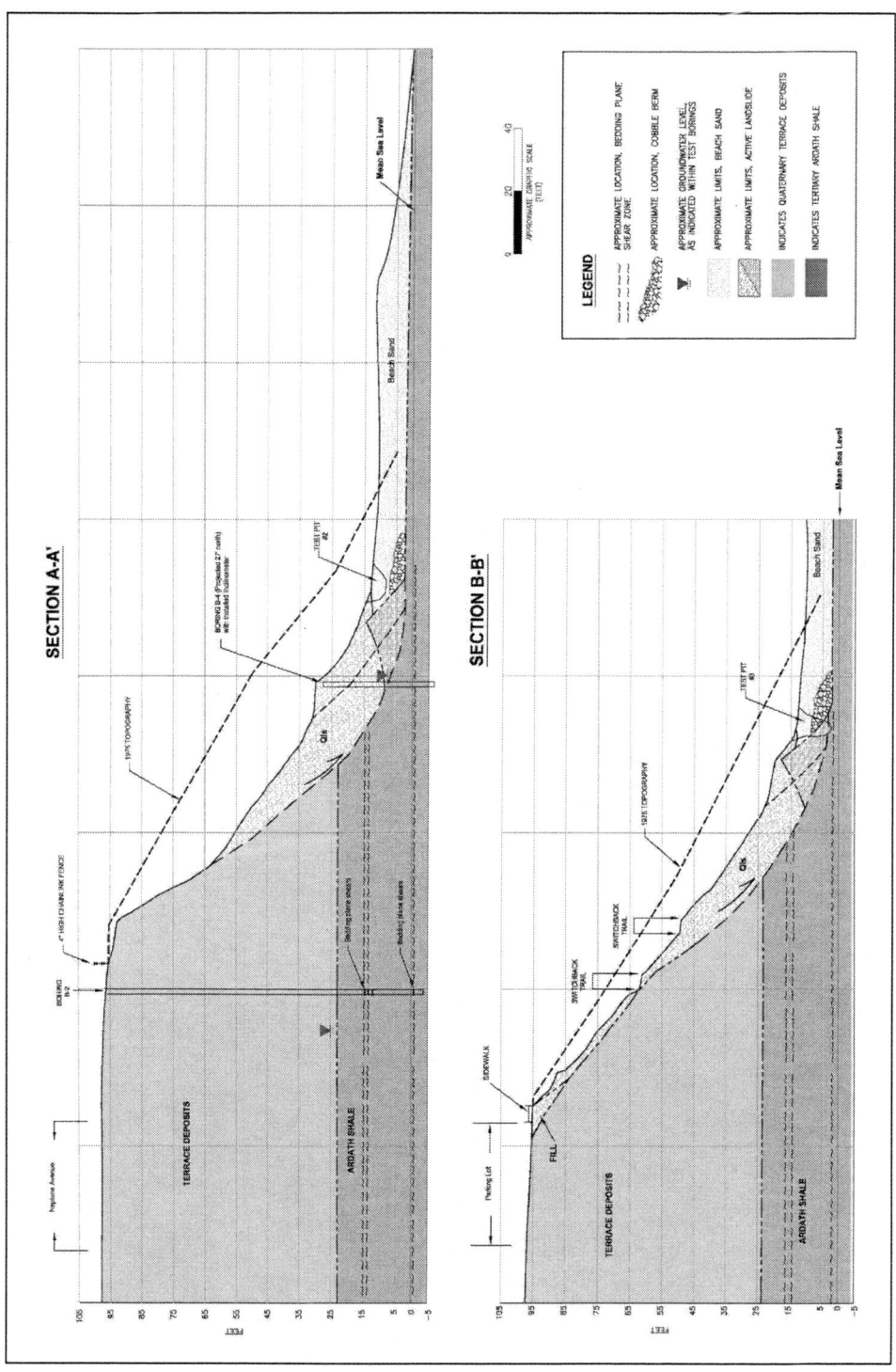

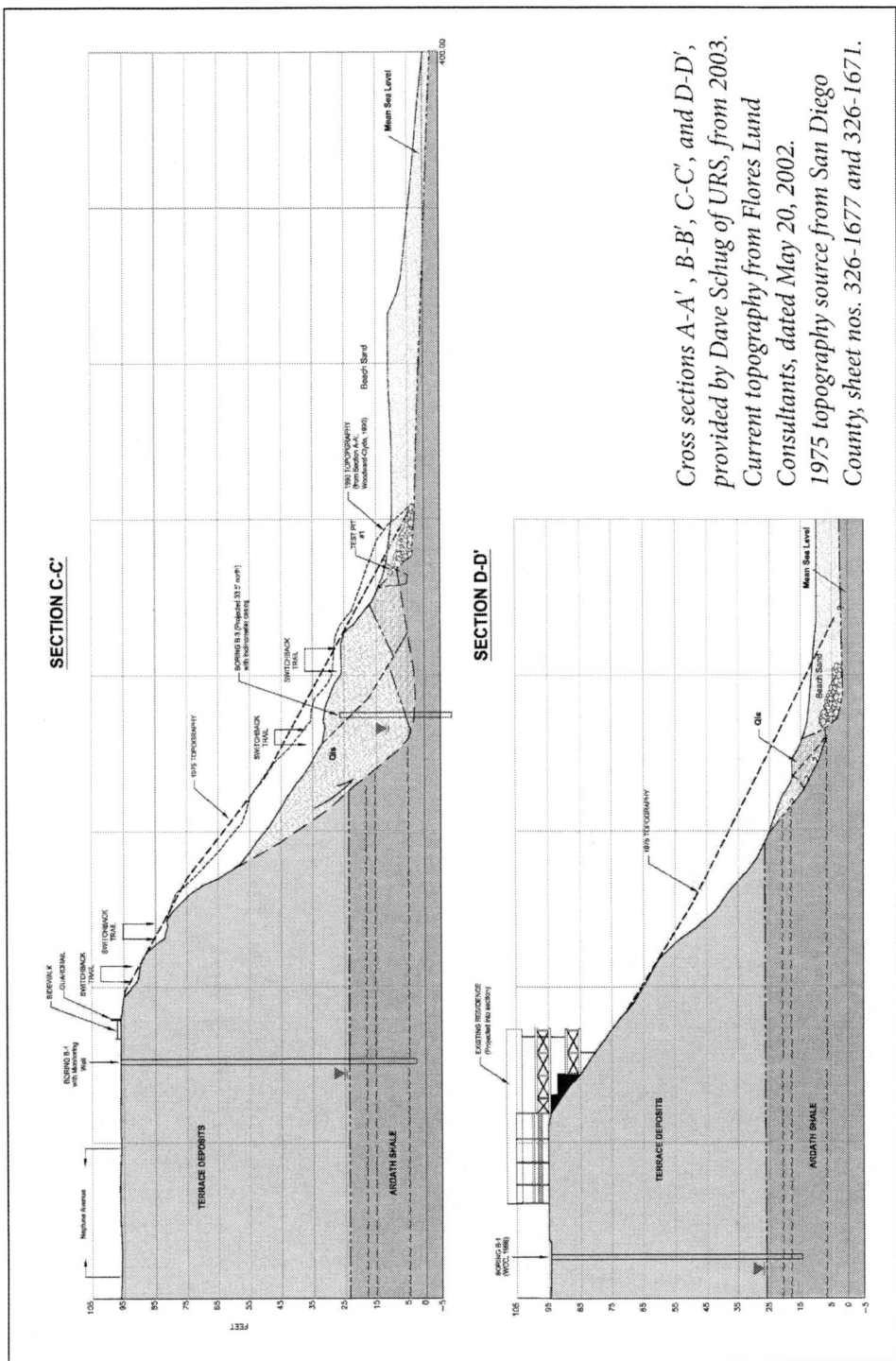

Cross sections A-A', B-B', C-C', and D-D', provided by Dave Schug of URS, from 2003. Current topography from Flores Lund Consultants, dated May 20, 2002. 1975 topography source from San Diego County, sheet nos. 326-1677 and 326-1671.

REFERENCES

Abbott, P.L., 1999: *The Rise and Fall of San Diego*, Sunbelt Publications.

Atwater, T.M.1998, Plate Tectonic History of Southern California with Emphasis on the Transverse Ranges and Northern Channel Islands: American Association of Petroleum Geologists, Pacific Section MP 45, p. 1–8.

Barberopoulou, Aggeliki, Mark R. Legg, Burak Uslu, Costas E. Synolakis, 2011, Reassessing the tsunami risk in major ports and harbors of California I: San Diego. Natural Hazards. vol. 58, p. 479–496.

Batiquitos Lagoon Foundation, 2012, History Timeline. http://www.batiquitosfoundation.org/newsite/naturewildlife_factsheets.php

Blake, W.P., 1855, Geological Map of the Country between San Diego and the Colorado River, California, scale 1:608,228, in Parke, J.G., Lt., and Campbell, A.H., 1854-1855, Report of Explorations for Railroad Routes from San Francisco Bay to Los Angeles, California, West of the Coast Range, and from The Pimas Villages on the Gila to the Rio Grande, near the 32d Parallel of North Latitude, in: Reports of Explorations and Surveys to Ascertain the Most Practicable and Economical Route for a Railroad from the Mississippi River to the Pacific Ocean, made under the direction of the Secretary of War, in 1853–6, According to Acts of Congress of Mar 3, 1853, May 31, 1854, and August 4, 1854, Volume VII., Washington, Chapter XVIII – Geology of the District from San Diego to Fort Yuma, 119–129, Chapter XXVIII – List of Minerals Collected, 187-188, Plate VIII.

Bushee, Jonathan, John Holden, Barbara Geyer and Gordon Gastil, 1963, Lead-Alpha Dates for some Basement Rocks of Southwestern California. Geological Society of America Bulletin, June 1963, v. 74, no. 6, p. 803–806.

California Energy Commission, 2012, Energy Commission Approves Power Plant in Carlsbad, News Release. May 31. http://www.energy.ca.gov.

California Geological Survey, 1965, Geologic Atlas of California Map No. 019 (Santa Ana Sheet), 1:250,000 scale. Compilation by: Thomas H. Rogers. Interactive Map: http://www.quake.ca.gov/gmaps/GAM/santaana/santaana.html. Reference 35: "San Diego State College student mapping under the direction R. G. Gastil, 1954–1964; includes Oceanside, San Luis Rey (also mapped by B. F. Jones, University of Southern California, M.S. thesis, 1959), Encinitas, San Marcos, Valley Center, Rancho Santa Fe, and Escondido 7-1/2' quadrangles (local additions and modifications from J.C. Taylor, Geologic map of parts of the Oceanside, San Luis Rey, and Morro Hill quadrangles, unpublished map, 1950, on file Pomona College)."

California Geological Survey, 2002, Regional Geologic Map No. 2, Geologic map of the Oceanside 30'×60' Quadrangle, California, 1:100,000 scale. Compiled by: Michael P. Kennedy and Siang S. Tan. Digital preparation by: Kelly R. Bovard, Rachel M. Alvarez, Michael J. Watson, and Carlos I. Gutierrez. http://www.quake.ca.gov/

gmaps/RGM/oceanside/oceanside.html Accessed: April 9, 2012.

Carlsbad Historical Society, 2006, Calavera Dam Construction, Carlsbad Time Lines, First Quarter, v. 8, no. 1, p. 3. http://carlsbadhistoricalsociety.com/Carlsbad%20 Historical%20Society_files/newsletter/newsletter6Q1.pdf

City of Oceanside, 2012, Request for Proposal Airport Master Plan, June 11. http://www.ci.oceanside.ca.us/gov/manager/propmgmt/opps/airportmp.asp

Dartnell, Peter, William R. Normark, Neal W. Driscoll, Jeffrey M. Babcock, James V. Gardner, Rikk G. Kvitek, and Pat J. Iampietro, 2007, Multibeam Bathymetry and Selected Perspective Views Offshore San Diego, California, U.S. Geological Survey Scientific Investigations Map 2959. http://pubs.usgs.gov/sim/2007/2959

Deméré, Thomas A., 2008, Volcanoes of San Diego County, San Diego Natural History: Field Notes, April 2008. http://www.sdnhm.org/archive/research/readings/fn_0408.html

Elliott, W.J., 1985, Geology of Morro Hill area, northwestern San Diego County, California in Abbott, P.L., editor, On the manner of deposition of the Eocene strata in northern San Diego County: San Diego, California, San Diego Association of Geologists, p. 85.

Elliott, W.J., 2001, Coastal Landsliding, Leucadia, California, in Stroh, R.C., editor, Coastal Processes and Engineering Geology of San Diego, California. San Diego Association of Geologists.

Engstrand, I.H.W., and Ward, M.F., 1995, Rancho Guajome, An architectural legacy preserved: San Diego Historical Soc. Quarterly, v. 41. San Diego History online, Univ. San Diego <www.sandiegohistory.org/journal/95fall/guajome.htm>

Everhart, D.L., 1951, Geology of the Cuyamaca Peak quadrangle, San Diego County, California: California Division of Mines and Geology Bulletin, no. 159, p. 51–115, (incl. geologic map, scale 1:62,500).

Four Directions Institute, Luiseno: <www.fourdir.com/luiseno.htm>

Fritsche A.E., Weigand P.W., Colburn I.P., Harma R.L., 2001, Transverse/Peninsular Range Connections-Evidence for the Incredible Miocene Rotation: Geologic Excursions in Southwestern California, SEPM (Society for Sedimentary Geology), Pacific Section, book 89, pg. 101–146.

Garrison, T. S. (2011). Essentials of Oceanography. (6th ed.). Canada: Brooks/Cole Pub. Co.

Geocon, 2012, Southwest Fisheries Science Center, La Jolla, California, Slope Mitigation For Building A, July 20. http://www.wpconstruction.com/current/lajolla/lajolla65/7%20-%20Geotechnical%20Report/Soils%20Reports/6Q%20Slope%20Mitigation%20For%20Bldg.%20A.pdf

Gibbons, H. (2011, July 14). Japan Lashed by Powerful Earthquake, Devastating Tsunami. Retrieved from http://soundwaves.usgs.gov/2011/03/

Griggs, G. B. (1992). California's Coastal Hazards. Journal of Coastal Research, 12, 1–15.

Hart, M.W., 1972, Landslides of West-Central San Diego County, California, M.S. Thesis, San Diego State University. Department of Geological Sciences. January 10. http://sdsu-dspace.calstate.edu/handle/10211.10/2146

Henry, Barbara, 2008, "Geologists hope to date Mount Calavera, Morro Hill's volcanoes": *North County Times*. June 23.

Kennedy, Michael P., 1975, Geology of the San Diego Metropolitan Area, California, Section A, Western San Diego Metropolitan Area, Del Mar, La Jolla, and Point Loma 7 1/2-minute quadrangles. Bulletin 200, California Division of Mines and Geology. http://archive.org/details/geologyofsandieg00kennrich

Kuhn, Gerald G., and Francis P. Shepard, 1984, *Sea Cliffs, Beaches, and Coastal Valleys of San Diego County: Some Amazing Histories and Some Horrifying Implications*. Berkeley: University of California Press. http://ark.cdlib.org/ark:/13030/ft0h4nb01z/

La Page, Steve, 2007, Potential Adverse Changes in Agua Hedionda Lagoon Resulting from Abandonment of the Lagoon Intake. May 18. http://www.waterboards.ca.gov/rwqcb9/press_room/announcements/carlsbad_desalination/updates_3_13_09/item_7_h.pdf

Larsen E.S. Jr. 1948, Batholith and Other Associated Rocks of Corona, Elsinore, and San Luis Rey Quadrangles, Southern California: The Geologic Society of America Memoir 29, 1 plate, scale 1:125,000.

Larsen, Jr., E. S., Everhart, D. L., and Merriam, R., 1951, Crystalline rocks of southwestern California: California Division of Mines Bulletin 159, 128 p.

Marvin, R.F., and Dobson, S.W., 1979, Radiometric ages; compilation "B", U.S. Geological Survey: Isochron/West, no. 26, p. 3–32. http://ngmdb.usgs.gov/Geolex/NewRefsmry/sumry_5457.html Accessed April 22, 2012.

McCurry, J. (2011, March 21). Japan Quake Death Toll Passes 18,000. Retrieved from http://www.guardian.co.uk/world/2011/mar/21/japan-earthquake-death-toll-18000.

Miller, F. S., 1937, Petrology of the San Marcos gabbro, southern California: Geological Society of America Bulletin, v. 48, p. 1397-1426.

Minch, J., E. Minch, and J. Minch, 1998, *Roadside Geology and Biology of Baja California*.

Miskwish, M.C., 2007, Kumeyaay, A History Textbook, v. 1, Precontact to 1893: Sycuan Press, El Cajon, CA, 173 p. Press, El Cajon, CA, 173 p.

Morton, P. K., R. V. Miller, and J. R. Evans. 1976. Environmental Geology of Orange County, California. California Division of Mines and Geology Open-File Report 79-08. 1:48,000.

National Oceanographic and Atmospheric Administration (NOAA), 2012a, National Geophysical Data Center, 1854 Marigram for San Diego. Climate Database Modernization Program, collection 1854 to 1981. http://www.ngdc.noaa.gov/hazard/tide.shtml

NOAA, 2012b, Tides and Currents, Tide Data, Station 9410230, La Jolla, California. http://tidesandcurrents.noaa.gov

NOAA, 2012c, Mean Sea Level Trend Station 9410230, La Jolla, California. http://tidesandcurrents.noaa.gov/sltrends/

NRG, 2012, Encina Power Plant. http://maps.nrgenergy.com/

Preserve Calavera, 2012, Calavera Area Information. http://www.preservecalavera.org/calavera.html

Project Clean Water, 2012, Watersheds: Santa Margarita Watershed. http://www.projectcleanwater.org

San Diego Association of Governments (SANDAG), 2012, Fact Sheet. Santa Margarita River Railroad Bridge Replacement and Second Track Project. July. http://www.sandag.org/uploads/publicationid/publicationid_1647_14259.pdf

Scott, E., 2004, Early records of extinct Bison from southern California and the beginning of the Rancholabrean North American Land Mammal Age: Southern California Academy of Sciences Annual Meeting, May 14–15, California State University, Long Beach.

Scripps Institution of Oceanography (SIO), 2012, http://sio.ucsd.edu/About/Venue_Rentals/Scripps_Pier/.

Sisson, Paul, 2012, San Onofre: Nuclear plant inspection continues after water leak, *North County Times*, February 6. http://www.nctimes.com/news/local/oceanside/san-onofre-nuclear-plant-inspection-continues-after-water-leak/article_8f583b27-7562-580d-8004-215525a94e8c.html

Southern California Edison (SCE), 2012, SONGS Fact Sheet. http://www.sce.com/PowerandEnvironment/PowerGeneration/SanOnofreNuclearGeneratingStation/songsfactsheet.htm

State of California, 2009, Tsunami Inundation Map for Emergency Planning, La Jolla Quadrangle, San Diego County; produced by California Emergency Management Agency, California Geological Survey, and University of Southern California – Tsunami Research Center; dated June 1, 2009, mapped at 1:24,000 scale.

Tan, S.S. 1986, Landslide Hazards in the Encinitas Quadrangle, San Diego County, California: California Division of Mines and Geology, open File Report 86-8, 3 plates, scale 1:24,000.

Tan, S.S. 1987, Landslide Hazards in the Rancho Sante Fe Quadrangle, San Diego County, California: California Division of Mines and Geology, open File Report 86-15, 3 plates, scale 1:24,000.

Tan, S.S., 1999c, Geologic map of the San Onofre Bluff 7.5' quadrangle, San Diego and Orange counties, California: A digital database: California Geological Survey preliminary geologic map website, http://www.conservation.ca.gov/cgs/rghm/rgm/preliminary_geologic_maps.htm.

Tan, S.S., 2001, Geologic Map of the Morro Hill 7.5' Quadrangle, San Diego County, California: A Digital Database. Version 1.0. Digital database by Kelly Corriea. ftp://ftp.consrv.ca.gov/pub/dmg/rgmp/Prelim_geo_pdf/morro_hill.pdf.

Tan, S.S., Kennedy M.P. 1996, Geologic Maps of the Northwestern Part of San Diego County: California Division of Mines and Geology, open File Report 96-02, 3 plates, scale 1:24,000.

Tarbuck and Lutgens 2007, Earth, An Introduction to Physical Geology, Prentice Hall.

Trujillo, A. P., & Thurman, H. V. (2014, in press). Essentials of Oceanography. (11th ed.). New Jersey: Pearson Education.

Turbeville, John, 2008, The Geology of Calavera Hills, North San Diego County, California. *Geoscience Investigations in Northern San Diego County and Beyond: Student-Directed Explorations.* http://www.miracosta.edu/home/jturbeville/Publications.html Accessed: April 20, 2012.

United States Nuclear Regulatory Commission, 2012, San Onofre Nuclear Generating Station, Unit 2. http://www.nrc.gov/info-finder/reactor/sano2.html

University of California, 2012, University of California History, San Diego: Historical Overview, derived mainly from Verne Stadtman's Centennial Record, published in 1968. http://sunsite.berkeley.edu/~ucalhist/general_history/campuses/ucsd/overview.html

USGS, 2009, Water-Data Report 2009, 11042000 San Luis Rey River at Oceanside, California. http://wdr.water.usgs.gov/wy2009/pdfs/11042000.2009.pdf

USGS, 2012a, National Geologic Map Database GEOLEX, Geologic Names Lexicon. http://ngmdb.usgs.gov/Geolex Accessed: April 18, 2012.

USGS, 2012b, Significant Earthquake Archive: Report Magnitude 9.0 - Near the East Coast of Honshu, Japan. Earthquake Hazards Program. http://earthquake.usgs.gov/earthquakes/eqinthenews/2011/usc0001xgp

Vista Irrigation District, 2012, Our History. http://www.vid-h2o.org/aboutus/ourhistory.asp

Weber, Jr., F. H., 1963, Geology and Mineral Resources of San Diego County, California. California Division of Mines and Geology County Report 3. http://www26.us.archive.org/details/geologyandminer03webe

Weigand P.W., Savage K.L., Nicholson C. 2002, The Conejo Volcanics and Other Miocene Volcanic Suites in Southwestern California: Geological Society of America Special Paper #365.

Wilson, K.L., 1972, Eocene and related geology of a portion of the San Luis Rey and Encinitas quadrangles, San Diego County, California: Unpublished M.A. thesis, University of California, Riverside. Masters thesis, 135 p., 2 plates, scale 1:24,000.

Calavera Lake, Carlsbad, February 17, 1984.
Photo: Woodrow L. Higdon, Geo-Tech Imagery.

Oblique aerial view of a 700 foot long by 320 foot wide landslide that occurred February 1978 at San Onofre State Beach near Stop 3, about a mile southeast of San Onofre Nuclear Generating Station March 10, 2008.
Photo: Woodrow L. Higdon, Geo-Tech Imagery.

II. Papers

The Great Wave off the Coast of Kanagawa,
wood-block print, by Hokusai.
Part of The Thirty-Six Views of Fuji *series (1823–29).*
Source: wikipedia.org.

PRELIMINARY SEARCH FOR ELTANIN IMPACT TSUNAMI DEPOSITS IN SOUTHERN CALIFORNIA:
A Possible Early Pleistocene Chronozone

Lawrence L. Busch
Brian J. Swanson
Janis L. Hernandez
Brian P.E. Olson
volcanmtn@gmail.com

PREFACE

This article is the result of an *ad hoc* collaboration for this guidebook. We present here an exercise (or adventure) in theoretical and applied geoscience. Our intent is to disseminate knowledge of a geologically significant Early Pleistocene event, the Eltanin bolide impact, in order to stimulate discussion and encourage further investigation. This event is well documented in magnitude, space, and time. Published modeling suggests it produced an unusually large tsunami that would reasonably be recorded in the stratigraphic record of southern California. We surveyed the geologic literature for descriptions of deposits with characteristics consistent with potential tsunami deposits and we performed limited reconnaissance-level field observations of promising sites. Based on this preliminary work, we identify stratigraphic sections and individual deposits as candidates for further investigation.

INTRODUCTION

About 2.5 million years ago, the 0.5- to 4-km-diameter Eltanin asteroid impacted into deep water of the Southern Ocean, 1400 km due west of Cape Horn, South America (Margolis and others, 1991; Gersonde and others, 1997; Frederichs and others, 2002; Gersonde and others, 2005; Kyte and others, 2005; Kyte and others, 2006). This impact produced a large tsunami wave-set. Modeling of the size and extent of this tsunami across the Pacific Ocean (Ward, 2002; Ward and Asphaug, 2002) indicates the open ocean wave-set may have been about 30 m high when it hit California, with an estimated wave-train duration of several hours (Figure 1). Concentration and refraction of the wave-set as it obliquely encountered the length of the Baja California peninsula, and shoaling effects upon bathymetric sounding, would likely have amplified these waves as they approached coastal southern California.

Since we know when this impact occurred, and that the resulting tsunami was likely larger than any earthquake-generated tsunami or storm waves, we seek to identify unique deposits from this event that may be preserved in the sedimentary record of southern California. If deposits from this event can be documented, they would represent a potentially widespread Early Pleistocene, geologically instantaneous, isochronous chronostratigraphic unit (a *chronozone*) on the margins of the Pacific Ocean. Recognition of such deposits could provide a critical Early Pleistocene time marker for a vast geographical area. The importance of identifying this chronozone gains added significance considering the recent revision of the Plio-Pleistocene boundary to 2.58 Ma (Gibbard and others, 2010). We identify several candidate deposits in the San Diego area, eastern Ventura Basin, and Los Angeles Basin for further investigation toward defining local examples of Eltanin impact tsunamiites.

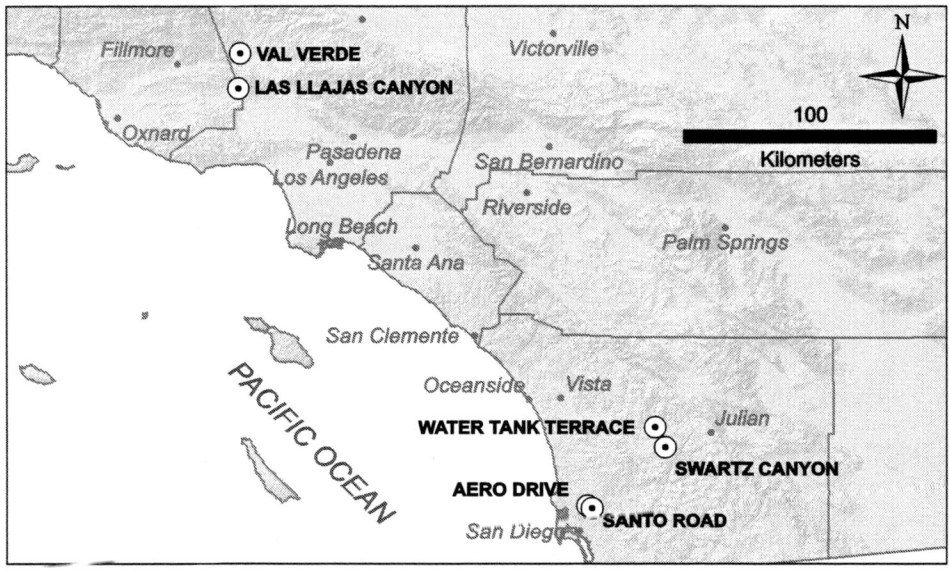

Map 1. *Locations of candidate tsunami deposit outcrops pictured in Figures 2 through 12.*

Evidence for an impact tsunami may include: sharp or erosional basal contacts; scour-featured erosion surfaces; boulder-filled channels, sheets, and aprons; chaotic bimodal sediment distribution; "floating" boulders in a sandy matrix; coarse-grained high-energy offshore return flow deposits; "dump deposits" — chaotic sediment mixture and minimal sorting indicating a short-duration depositional process; inland sand sheets blanketing existing topography; rip-up clasts; mud-capped deposits in muddy environments; mud cracks; marine fossils in a terrestrial environment; fossil hash beds; mixed marine/terrestrial fossil beds; presence of microtektites, iridium anomalies; etc. Optimal environments for the preservation of tsunami deposits include bays, estuaries, lagoons, tidal marshes, low-lying terraces and floodplains, and shallow marine shelf environments below typical wave base (Scheffers and Kelletat, 2004; Rhodes and others, 2006; Peters and Jaffe, 2010; De Martini and others, 2011; Varela and others, 2011).

SAN DIEGO REGION

We identify three stratigraphic sections with potential to contain a sedimentary record of the Eltanin impact tsunami: 1) near the top of the Pliocene San Diego Formation; 2) near the base of the Plio-Pleistocene Lindavista Formation; and 3) inland sandstone deposits previously interpreted as terrestrial fan deposits. We note that the local Plio-Pleistocene relative sea level was likely several hundred meters above current sea level (Hertlein and Grant, 1944). Peterson (1970) suggests that the marine Poway Terrace surface that increases in elevation from 800 feet in the west near Miramar, to 1200 feet in the east near State Highway 67 may have been wave-cut as recently as early Pleistocene. The extensive San Diego Terrace was already in existence by Pliocene time based on the presence of the San Diego Formation in that stratigraphic position south of the San Diego River.

San Diego Formation. The upper Pliocene San Diego Formation (Hertlein and Grant, 1944; Kennedy and Peterson, 1975; Kennedy, 1977) is a primary candidate section for evidence of the Eltanin impact tsunami event. A 30 meter-thick boulder conglomerate near the International Border is one candidate deposit in this section:

"about one mile west of Boundary School, a considerable thickness of conglomerates and coarse indurated sands is exposed…This conglomerate stratum is fully 100 feet thick, and the large boulders range up to three feet in maximum diameter" (Hertlein and Grant, 1944, p. 51).

In an impact tsunami model this deposit may represent tsunami return flow channel deposits offshore of the ancestral Tijuana River.

Another potential candidate deposit assigned to the San Diego Formation by Hertlein and Grant (1944, p. 51) is an unusually resistant, angular coarse-grained sandstone occurring as "a prominent outcrop of very coarse-grained gravelly sandstone about three and a quarter miles west of Sweetwater Reservoir." Cleveland (1960, p. 7) reports a similar bed in the San Diego Formation in the Otay River area as "well-bedded, resistant to weathering, and about 10 feet thick, composed of limonite-cemented quartz fragments and angular clasts of pebble- to cobble-sized volcanic rocks." According to Kuper and Gastil (1977), these resistant, gravelly coarse sands are part of the distal, gritstone facies of the Sweetwater Formation (Artim and Pinckney, 1973; Scheidemann and Kuper, 1979) of pre-late Oligocene age (Kennedy and Tan, 2008). The Sweetwater Formation is subdivided into four facies: an angular [boulder] conglomerate facies; a gritstone facies (Walsh and Deméré, 1991, place these two beds in the Otay Formation); a mudstone facies; and a sandstone facies. For the purpose of this paper, the proximal angular conglomerate facies, and the distal gritstone facies are of primary interest; Kuper and Gastil, (1977) describe these:

"The gritstone facies is found as lenses in the angular conglomerate and as massive beds up to 16 meters thick…The color ranges from bright white to tan. It is a lithic arkose in composition, with half-centimeter grains of quartz, feldspar and rock chips being the primary components. Within the Paradise Valley area, this facies is in part well cemented with silica and forms resistant exposures. A particularly resistant exposure just north of Paradise Valley Road was pictured by Hertlein and Grant (1944). Sediments within this facies are massive and poorly sorted and resemble sheetflood deposits"… "The angular conglomerate facies contains detritus that ranges in size from pebbles and sand matrix to giant boulders, which are angular to subrounded. The clasts consist largely of weakly metamorphosed volcaniclastic of the adjacent Santiago Peak Formation, with lesser amounts of granodiorite and gabbro. Locally, as southeast of Jamul, the matrix is poorly sorted, clay-rich, and in some places iron oxide stained.

The most extensive exposures are found in Proctor and Otay Valleys... The most easterly exposures are found capping ridges 8 km southeast of Jamul (approximately 350 to 400 meters above sea level)" (p. 11).

Deposits so described record a large-scale, high energy erosion and deposition event consistent with sheet flood deposits described by Peters and Jaffe (2010), and "bimodal" tsunami deposits of Scheffers and Kelletat (2004). If further investigation demonstrates that these deposits are actually Plio-Pleistocene age, then they may represent Eltanin tsunami return flow channel and distal sheet deposits offshore of the ancestral Otay and Sweetwater Rivers. However, if these bimodal angular boulder conglomerates are part of the pre-late Oligocene Sweetwater Formation, then they may record a large, unrecognized, pre-late Oligocene, post-late Eocene tsunami event.

Lindavista Formation. Selected basal beds of the reddish-brown conglomeratic sandstone marine terrace deposits of the Plio-Pleistocene Lindavista Formation (Hanna, 1926; Moore, 1972; Kennedy, 1973; Kennedy and Peterson, 1975) represent a second set of candidate deposits. Hertlein and Grant (1944), in describing the Lindavista Formation ("Sweitzer formation" of Hertlein, 1929), report:

"Wherever this formation occurs on the mesa it tends to produce steep-walled valleys with angular shoulders. A peculiar feature connected with its origin, is its unbroken extension from the top of the principal terrace, the San Diego Mesa, over the intervening slope to the next lower terrace. This can be seen in but a few places (for example in the Chula Vista region) but it must be considered as having an important bearing on its origin" (p. 19) ... "At a number of places it not only caps the higher mesas but extends down over their margins onto the next lower terrace and apparently in some instances continues even lower" (p. 63).

In these passages the authors seem to suggest the possibility of an unusual origin for these deposits relative to normal (uniformist) geological processes. For example, how could an ongoing geological process move many millions of cobbles across a wide area and form thin, uniform deposits over existing multi-terraced topography? In an impact tsunami scenario, these deposits may represent rip-up clasts transported and redeposited in the turbulent, high-energy, short-duration environment in the littoral zone.

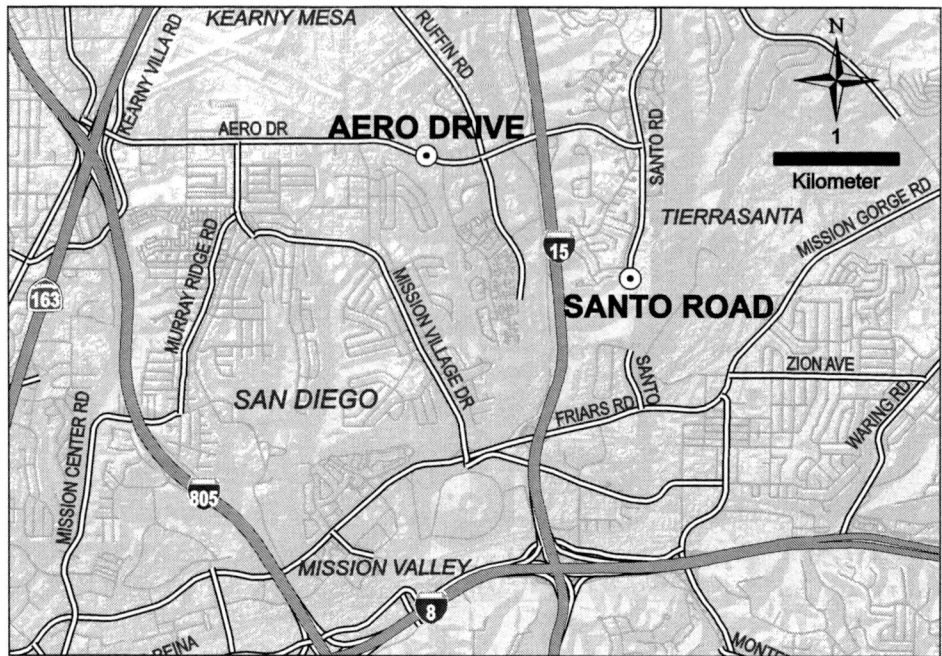

Map 2. Locations of Lindavista Formation outcrops in San Diego pictured in Figures 2 through 4.

A roadcut exposure at the southern terminus of Santo Road in the Tierrasanta neighborhood of San Diego mapped as Lindavista Formation (Kennedy and Peterson, 1975) and Qvop-8 (very old paralic deposits - unit 8; of middle to early Pleistocene age) by Kennedy and Tan (2008), displays a record of sediment deposited in a turbulent, high-energy environment of short duration (Figure 2) consistent with tsunami "dump deposits" described by Scheffers and Kelletat (2004). Here the basal unit consists of a one- to two-meter-thick bed of chaotic, bimodal, upwardly-fining matrix-supported, muddy cobble conglomerate. Clasts include Poway-type metavolcanic rocks and quartzite, local meta-andesite, deeply weathered granite, and possible poorly preserved mudstone rip-up clasts. The lower contact is a clean, sharp, fresh, current-swept erosion surface cut into a clean white sandstone facies of the Stadium Conglomerate. Preserved burrows in the upper surface of the underlying sandstone (Figure 3) indicate that this location did not experience significant, deep erosion immediately prior to deposition of the overlying chaotic basal conglomerate. The presence of burrows and the absence of any soil profile suggest a marine environment. The burrows are filled with pale reddish-orange, iron oxide-stained sandstone and pebbles of the overlying chaotic basal conglomerate — demonstrating a high-energy depositional environment of very short duration. These deposits are similar in stratigraphic position, lithology, structure, and scale to deposits in the Aero Drive roadcut at Montgomery Field approximately 3 km to the west (Figure 4). This muddy, chaotic, matrix-supported conglomeratic character is not consistent with the expectation of winnowed, well sorted deposits that would result from cyclic or ongoing processes of erosion and deposition expected in beach, surf, tidal, or near-shore environments.

The deposits at Santo Road, mapped as Pleistocene Lindavista Formation and Qvop-8, are 2.5 km north of exposures of the Pliocene San Diego Formation, which occur at the same stratigraphic position; at the same (or similar) elevation; and on the same (or similar) marine terrace. About 2 km of the intervening sedimentary record has been removed by the San Diego River. Since the San Diego Formation was necessarily deposited upon an existing marine terrace in the Pliocene, it would not be implausible if (1) sediments of nearly the same age were deposited on the same marine terrace; (2) at the same (or similar) elevation; (3) in the same (or similar) stratigraphic position; (4) parallel to the Plio-Pleistocene shoreline, at a distance of 2.5 km to the north.

Introducing additional complexity into the evaluation of a speculative impact tsunami model for basal deposits of the Lindavista section north of Mission Valley, Kern and Rockwell (1992) and Kennedy and Tan (2008) identified and mapped a series of at least 16 middle to late Quaternary emergent marine terraces in the San Diego region. Using correlations between uranium series ages of corals and amino acid "dates," and assuming a constant uplift rate of 0.13 m/ka, these authors date this series of beach ridges and terraces deposits — including area previously mapped as Lindavista Formation — back to about 1.6 Ma. While it seems implausible that a record of a 2.5-Ma event would be found in a section mapped as less than 1.6 Ma, we are not convinced that the age of the basal Lindavista deposits at the Santo Road site must necessarily be the same age as the overlying beach deposits. For example: (1) Hertlein and Grant (1944, p. 64) considered the beach ridges of the Lindavista Formation to be younger than the underlying sediments; (2) immediately south of

Mission Valley, Pleistocene marine terrace abrasion has not removed Pliocene age sediment of the San Diego Formation from the same abrasion surface 2 km south of the Santo Road site; and (3) the character of basal deposit at the Santo Road site is not consistent with the erosional and depositional processes expected in a near-shore environment. Since it is not clear that ages applied to Quaternary beach deposits by Kern and Rockwell (1992) and Kennedy and Tan (2008) necessarily apply to the underlying chaotic basal conglomerates observed at the Santo Road site, we concede that the stratigraphic position and sedimentary record at the Santo Road location are potentially consistent with a model for deposition by Eltanin impact tsunami action.

Flood Deposited Sandstone of Ramona. A third group of candidate deposits in the San Diego region are represented by a widely occurring but poorly exposed unnamed muddy sandstone unit mapped as QTf in the Ramona and San Pasqual 7.5' quadrangles (Todd and others, 2006; Hernandez and others, 2007), newly renamed as QTs in the most recent versions of both the Ramona and San Pasqual 7.5' quadrangles (in press). Similar deposits are observed from Valley Center south to Potrero, near

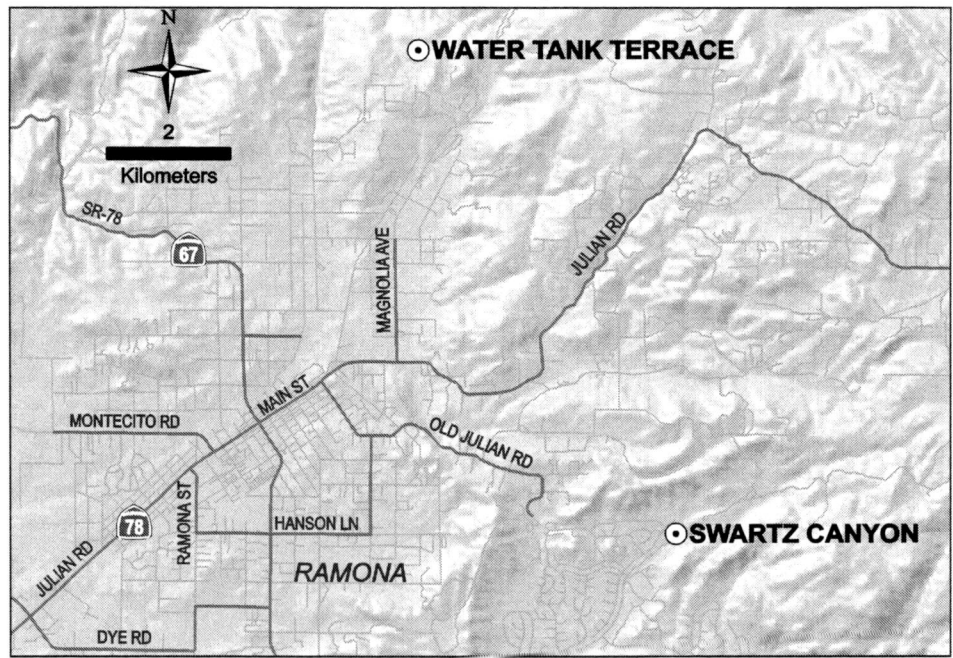

Map 3. *Locations of unnamed sandstone outcrops in Ramona pictured in Figures 5 through 8.*

the U.S.-Mexico border. These deposits occur as sets of one to three 0.3- to 0.6-meter-thick beds deposited on irregular, old terrace surfaces (Figures 5 and 6), as canyon fill, and as one- to two-meter-thick lobate deposits protruding from small drainages (Figure 7). Lobate deposits carry locally derived Poway-type and granitic clasts (Figure 8) up to one meter in diameter. Bedded deposits are often exposed in areas of badlands topography. QTs is poorly constrained stratigraphically, being deposited upon an uneven erosion surface on Cretaceous crystalline basement and overlain by Quaternary colluvium (Figures 5 and 6).

Outcrop descriptions QTs from typical locations within the Ramona 7.5' quadrangle include moderately to well-indurated, thin- to medium-bedded, silty, fine- to medium-grained sandstone. Matrix-supported angular (0.5 to 3 cm) fragments of locally derived pegmatitic rock are common. QTs contains rare Poway-type metavolcanic and quartzite cobble clasts up to 7 cm, and rounded, "rotten" granitic cobbles and sporadic boulders. Bedded sandstone deposits generally consist of three units; however, not all three are observed at all mapped locations. The upper silty sandstone unit is yellowish-brown (10YR 5/4), displays characteristic 10- to 30-cm-diameter polygonal desiccation cracks, distinctive clay- or silicate-cemented crude upwardly fining graded beds, and a cemented clay-rich layer at the base of the upper unit. Occasional fossilized silicate-lined root casts terminate in the basal clay layer. The upper unit is present in all mapped locations and may contain rare transported boulders (Figure 8). The middle sandstone unit is brown (7.5YR 5/4), highly mottled, commonly with ferruginous nodules, upwardly fining graded bedding, and generally coarser sand and pebble clast size than the upper unit. The lower gravelly sandstone unit is brown (7.5YR 5/4), mottled, contains a larger percentage of angular clasts, and is massively bedded. Lobate debris flow deposits are generally found at base of slopes and in small drainages. The mud-cracked upper silty sandstone and middle sandstone units are common in the lobate type of deposit.

The occurrence and appearance of QTs support the interpretation that it is a flood deposit. One possible interpretation is that these are local flash flood deposits. Another possible interpretation is that these deposits represent late-stage flood deposits of the Tertiary Ballena River system. A third possible interpretation is that these deposits represent indurated remnants of a tsunami sheet flood and return flow deposit consisting of a transported mixture of sand and material derived primarily from weathered surficial deposits of the Peninsular Ranges batholith. Due to the elevation and distance to the coast, the tsunami interpretation is considered speculative; but calculations that remove one-degree of regional uplift of the Peninsular Ranges from a hinge in coastal San Diego place these deposits near the Plio-Pleistocene sea level.

Compensating for the post-Pliocene change in relative sea level and regional uplift of the Peninsular Ranges may be useful in determining whether deposits mapped as QTs could have been deposited by a tsunami wave-set generated by the Eltanin impact. According to Hertlein and Grant (1944, p. 29), during the deposition of the Pliocene sediments the sea must have had a position far above the present mesa surface, possibly at an elevation of 800 feet or more above present sea level, or 300 or 400 feet above the present mesa level. Thus the San Diego Formation of Pliocene age must have once extended far north of Mission Valley and possibly considerably east of its present limits. Hypothetically, Plio-Pleistocene relative sea level corrected for generalized uplift of the Peninsular Ranges may be calculated as follows:

Given a right triangle with a 32 km base (the approximate distance from the easternmost Pliocene marine sediments to the Ramona area), and a 1-degree adjacent angle (hypothetical generalized regional angular uplift of the Peninsular Ranges), the height of the leg opposite the 1-degree angle (the amount of post-Pliocene uplift 32 km east of Pliocene marine sediments) is solved by the formula:

$$a = b\,(\tan A)$$

where,

a (uplift) = **b** (length of base/32 km) **tan A** (dip of beds/1-degree west)
 = 550 m of uplift 32 km east of coastal San Diego.

These 550 meters, plus an additional 250–350 meters of relative sea level elevation change from Plio-Pleistocene levels, suggests that it is well within the realm of possibility that the Ramona area in Plio-Pleistocene time was near enough to the coast in distance and elevation to be in a position to receive hypothetical Eltanin impact tsunami sediments.

VENTURA BASIN

Strata in the eastern Ventura Basin provide a generally continuous record of deposition during Late Pliocene and Early Pleistocene times. We identify several sites containing potential tsunami deposits in northwestern Los Angeles and eastern Ventura Counties within strata previously mapped in the upper part of the Pico Formation near the regressive transition to the overlying nonmarine Saugus Formation. These strata have been assigned a Late Pliocene age (prior to the recent revision of the age of the Plio-Pleistocene boundary) based on paleontological studies (Squires and others, 2006) and have been interpreted to represent deposition in shallow marine to transitional paleoenvironments, which are considered favorable for the accumulation and preservation of tsunami deposits. The specific lithologic units considered as candidate tsunami deposits consist of fossil hash beds composed primarily of oyster and pecten shell fragments (Figure 9). Thick, resistant, ridge-forming hash beds were observed at two locations, one on the south flank of the Santa Susana Mountains in Las Llajas Canyon (Santa Susana quadrangle) along the "Happy Camp" syncline (per Dibblee, 1992), and the other north of

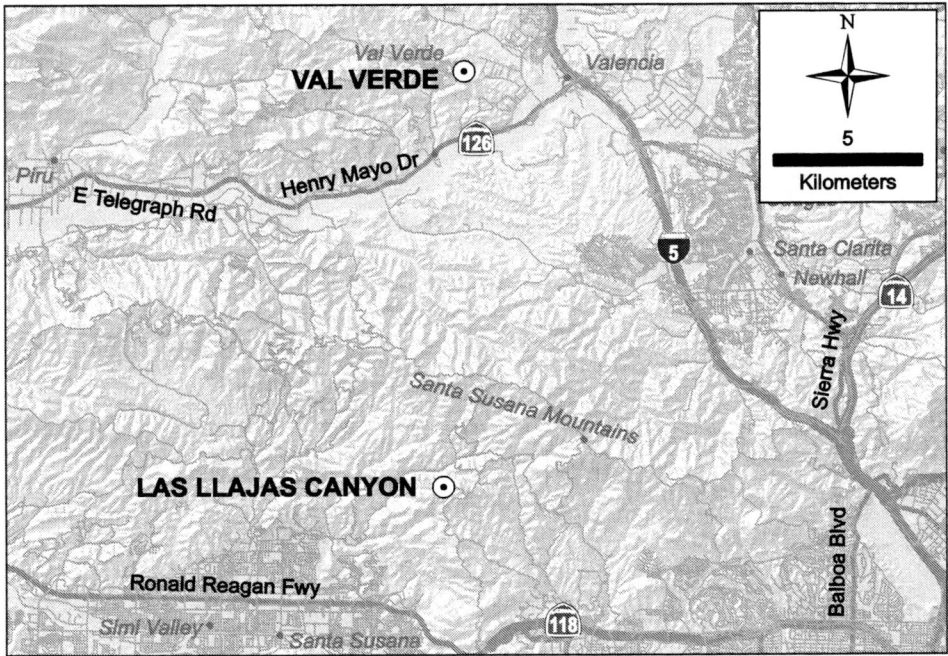

Map 4. Locations of fossil hash outcrops in upper Pico Formation in eastern Ventura Basin pictured in Figures 9 through 12.

the Santa Clara River southeast of the Town of Val Verde (Val Verde quadrangle; Dibblee, 1993). Multiple similar fossil hash beds have been reported at roughly the same stratigraphic position at other locations along the southern flank of the Santa Susana Mountains (Groves, 1991). Paleogeographic studies indicate the Val Verde strata were deposited near the north margin of the Ventura Basin and the Las Llajas strata were deposited near the south margin of the basin (Dibblee, 1995).

Laterally extensive shell hash beds are commonly interpreted to represent deposition in high energy environments as a result of wave action, storm surges, flooding events or tsunamis. The hash beds described above are considered candidate deposits from the Eltanin impact tsunami based on the following criteria:

1) The age of the strata is reported to be roughly coeval with the impact event.

2) The stratigraphic sequences at the two sites are indicative of deposition in shallow marine (Las Llajas) to transitional (Val Verde) depositional environments, which are considered favorable sites for the preservation of deposits resulting from tsunamis (Peters and Jaffe, 2010).

3) The observed deposits are thick (up to 2 m or more) and massive (Figure 9), which suggests deposition during a single large event.

4) The observed thickness of the hash beds is significantly greater than deposits documented from historic earthquake-generated tsunamis (Peters and Jaffe, 2010), which suggest a larger causal event.

5) Based on preliminary observations, the hash beds appear to be composed primarily of platy fragments of oyster and pecten shells, with very few intact individuals and none in apparent life position. Oysters in particular typically live in relatively quiet water environments, such as bays and lagoons, so the thickness of the deposits and broken nature of the shells suggests a high energy event and significant transport.

6) The observed beds exhibit chaotic structure of the platy shell fragments and a silty sand matrix (Figure 10), which suggest turbulent flow and rapid deposition. Deposition or reworking of platy shells by repetitive wave action would be expected to produce better sorting

and bedding-parallel or cross-bedded shell fragments (see Figure 11). Preservation of this chaotic structure indicates the depositional environment was therefore either above or below typical wave base or in a protected lagoon or bay, and that the deposits were not significantly reworked by subsequent periodic flood, storm or earthquake-generated tsunami events.

7) Based on preliminary observations, the deposits appear to display sharp upper and lower contacts with no distinct grading (see Figure 9), traits that are consistent with documented tsunami deposits (Peters and Jaffe, 2010).

The fossil hash bed near Val Verde occurs stratigraphically above a paleosol, which is typical of the interfingering nature of the Pico/Saugus Formation contact in this area and supports the depositional environment of the hash bed as being near the shoreline. However, the occurrence of this unit above the paleosol also suggests the possibility that it may have been deposited on land, and its singular occurrence in the section suggests the hash bed was the result of a rare large event. One reasonable scenario would be that the inflow tsunami wave(s) stripped and fragmented shells from a nearby (seaward) estuary, lagoon or bay and deposited them on a nearby floodplain, where the hash bed was locally preserved by subsequent fluvial deposition.

Multiple fossil hash beds observed in the upper Pico Formation at Las Llajas Canyon were interpreted by White (1983) to be the result of storm activity. Field observations suggest the presence of an angular discordance within the Pico Formation at the base of one of the more prominent fossil hash beds (Figure 12). This suggests a significant, intraformational scour event, which would be reasonably consistent with the effects a large tsunami could have on unconsolidated sediments in a shallow shelf environment. One reasonable scenario would be that scour occurred during the inflow of the tsunami wave and the shells were subsequently stripped from a nearby estuary, lagoon or bay, fragmented, and then deposited offshore below typical wave base during the return flow of the tsunami. Preliminary observation of additional hash beds observed higher in the section east of La Llajas Canyon indicates that some are bedded (see Figure 11), which suggests deposition by more traditional processes. Other hash beds that are more chaotic may represent deposition resulting from storm, flood or earthquake-generated tsunami events, which would be expected to occur relatively frequently in a geologic time frame (Varela and others, 2011). More detailed studies are needed to assess the origin of the various hash bed deposits at Las Llajas Canyon.

An additional candidate section in the eastern Ventura Basin that may contain Early Pleistocene tsunami deposits is the Sunshine Ranch Member of the lower Saugus Formation (Saul and Wootton, 1983). This unit was deposited in brackish water to near-shore environments that would have been favorable sites for the deposition and preservation of tsunami deposits. A specific candidate deposit was not assessed in this study, but one unit of potential interest is a locally extensive clay unit (slide plane referenced by Francuch, 2002) known informally as the "Friendly Valley Horizon" that occurs near the top of the section. This unit could represent more landward deposition as the wave energy attenuated and mud from a lagoon or estuary was stripped and redeposited by the tsunami wave.

LOS ANGELES BASIN AND VICINITY

Other candidate sections that may record evidence of the Eltanin impact tsunami may exist on the margins of the Plio-Pleistocene Santa Monica, San Gabriel and Channel Islands uplifts. Paleoenvironmental studies of the Upper Pliocene to Lower Pleistocene Series in the Los Angeles area indicate the paleoshoreline trended approximately east-west along the southern margin of the emerging San Gabriel Mountains, in what is now the San Gabriel Valley. A shallow marine facies of the Pliocene upper Fernando Formation overlain by nonmarine Pleistocene conglomerate is identified in industry well logs within the valley. These sediments are the most likely candidates to record evidence of this tsunami event north of the Los Angeles Basin. North of the Whittier Fault, the Fernando Formation is almost entirely buried under the valley alluvium. Just west of Whittier Narrows and south of the Whittier Fault, the upper Fernando Formation is exposed in both limbs of the Montebello Hills anticline. These sediments were likely deposited in a deeper portion of the Late Pliocene marine basin but may still contain discernible tsunami-related deposits. The remainder of the Fernando Formation that crops out south of Montebello in the Los Angeles

Basin (i.e., Palos Verdes Peninsula, western Puente Hills, and Newport Back Bay) are unlikely to contain readily discernible megascopic tsunami deposits from this event as these sediments were deposited in a deep marine basin; however, microscopic or chemical evidence of the impact itself (i.e., microtektites or iridium anomaly) may be detectable (Margolis and others, 1991; Peng, 1994; Gersonde and others, 2005; Kyte and others, 2005; Kyte and others, 2006).

CONCLUSIONS

Recognition of deposits from a geologically instantaneous impact event in the stratigraphic record could provide a critical Early Pleistocene time marker for a vast geographical area bounding the Pacific Ocean. The importance of identifying this chronozone gains added significance with the recent revision of the Plio-Pleistocene boundary to 2.58 Ma (Gibbard and others, 2010).

We identify stratigraphic sections in southern California where the Eltanin impact tsunami may be recorded, and identify selected candidate deposits at the Santo Road, Ramona, and Ventura Basin sites for further investigation. Additional studies are needed to establish whether the candidate sites discussed herein, or other sites, are the result of the Eltanin impact event. A detailed paleomagnetic study of upper Pliocene strata with a near-continuous record of deposition, such as the upper Pico Formation discussed herein, could provide a basis for assigning candidate deposits to the Eltanin impact event.

Unexpectedly, our literature survey of the sedimentary record for the San Diego area revealed deposits with characteristics suggesting that they may result from a separate large tsunami event. If the proximal angular boulder conglomerate facies and the distal gritstone facies of Sweetwater Formation are not of Plio-Pleistocene age, then these deposits may have originated from an unrecognized Pacific Ocean impact and tsunami event that occurred in the pre-late Oligocene to post-late Eocene time interval.

ILLUSTRATIONS

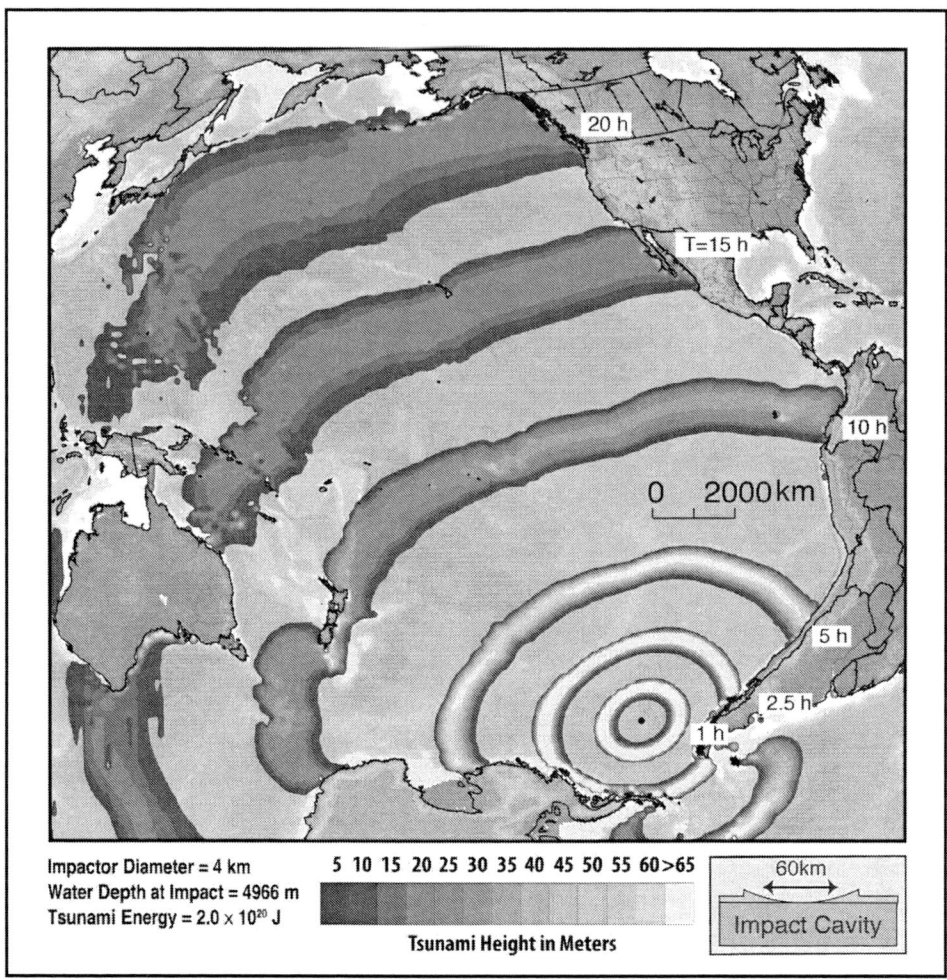

Figure 1. *Model of tsunami path and height for hypothetical 4 kilometer-diameter Eltanin impactor (Ward, 2002).*

Figure 2. *Chaotic basal conglomerate of Lindavista Formation deposited on sandstone facies of the Stadium Conglomerate at Santo Road site.*

Figure 3. *Chaotic basal conglomerate of Lindavista Formation filling burrows in sandstone facies of the Stadium Conglomerate at Santo Road site.*

Figure 4. *Basal conglomerate of Lindavista Formation deposited on Mission Valley Formation, Aero Drive road cut at Montgomery Field.*

Figure 5. *Bedded sandstone deposit (QTs), Swartz Canyon site, Ramona.*

Figure 6. *Bedded sandstone deposit (QTs), Water Tank Terrace site, Ramona. Note polygonal cracked layer as upper bed. Clay rich layer at top end of rock hammer contains fossilized root casts.*

Figure 7. *Lobate sandstone deposit (QTs), Water Tank Terrace site, Ramona. Polygonal cracked uppermost layer common, with graded bedding and occasional coarse pebbles in lower bed.*

Figure 8. *Lobate sandstone deposit (QTs) with "floating" boulder imbedded in surface of upper bed, Water Tank Terrace site, Ramona. Boulder is local plutonic basement.*

Figure 9. *Exposure of thick, massive fossil hash bed at Val Verde site.*

Figure 10. *Detail of chaotic structure in fossil hash bed at Val Verde site.*

Figure 11. *Detail of internally bedded fossil hash unit exposed in road cut on ridge east of Las Llajas Canyon.*

Figure 12. *View southwest of apparent angular discordance between fossil hash bed and underlying beds exposed on ridge east of Las Llajas Canyon. All exposed beds are mapped as Pico Formation per Dibblee (1992).*

REFERENCES

Artim, E.R., and Pinckney, C.J., 1973, La Nacion fault system, San Diego, California, *Geological Society of America Bulletin*, v. 84, no. 3, p. 1075-1080.

Cleveland, G.B., 1960, Geology of the Otay Bentonite Deposit, San Diego County California: California Division of Mines Special Report 64, 16 p., 1 Plate.

De Martini, P.M., Smedile, A., and Pantosti, D., 2011, Combining inland and offshore paleotsunami evidence: the Augusta Bay case study, *in* Brand, F., ed. Marine geo-hazards in the Mediterranean – CIESM workshop Monograph, 192 p.

Dibblee, T. W., Jr., 1992, Geologic Map of the Santa Susana Quadrangle: Dibblee Geological Foundation Map #DF-38, Map scale 1:24,000.

Dibblee, T. W., Jr., 1993, Geologic Map of the Val Verde Quadrangle: Dibblee Geological Foundation Map #DF-50, Map scale 1:24,000.

Dibblee, T. W., Jr., 1995, Tectonic and depositional environment of the middle and upper Cenozoic sequences of the coastal southern California region: *in* Fritsche, A. E., ed., Cenozoic Paleogeography of the Western United States – II: Pacific Section, SEPM Book 75, p.212-245.

Francuch, D. G., 2002, Stabilization of a portion of the Honby landslide complex, Santa Clarita, California: Geological Society of America Abstracts with Programs, v. 34, no. 5, p. A-94.

Frederichs, T., Bleil, U., Gersonde, R., and Kuhn, G., 2002, Revised age of the Eltanin Impact in the Southern Ocean, *Abstract* #OS22C-0286- American Geophysical Union, Fall Meeting. http://adsabs.harvard.edu/abs/2002AGUFMOS22C0286F

Gersonde, R., Kyte, F.T., Bleil, U., Diekmann, B., Flores, J.A., Gohl, K., Grahl, G., Hagen, R., Kuhn, G., Sierro, F.J., Volker, D., Abelmann, A., and Bostwick, J.A., 1997, Geological record and reconstruction of the late Pliocene impact of the Eltanin asteroid into the Southern Ocean: Nature v. 390, p. 357-363. http://epic.awi.de/Publications/Ger1997i.pdf.

Gersonde, R., Kyte, F.T., Frederichs, T., Bleil, U., Schenke, H.-W., and Kuhn, G., 2005, Late Pliocene impact of the Eltanin asteroid into the Southern Ocean – documentation and environmental consequences: Geophysical Research Abstracts; European Geosciences Union. http://www.cosis.net/abstracts/EGU05/02449/EGU05-J-02449.pdf.

Gibbard, P.L., Head, M.J., Walker, M.J.C., and Subcommission on Quaternary Stratigraphy, 2010, Formal ratification of the Quaternary System/Period and the Pleistocene Series/Epoch with a base at 2.58 Ma: Journal of Quaternary Science, v. 25, p. 96-102.

Groves, L., 1991, Paleontology and biostratigraphy of the Plio-Pleistocene lower Saugus Formation, Santa Susana Mountains, southern California: Unpublished Masters Thesis, California State University, Northridge, 372 p.

Hanna, M.A., 1926, Geology of the La Jolla quadrangle, California: University of California Publications in Geological Sciences Bulletin, v. 16, no. 7, p. 187–246, (incl. geologic map, scale 1:62,500).

Hernandez, J.L., Todd, V.R, Busch, L.L., and Tan, S.S., 2007, Geologic Map of the San Pasqual 7.5' Quadrangle, San Diego County, California: A Digital Database: California Geological Survey (1:24,000). ftp://ftp.consrv.ca.gov/pub/dmg/rgmp/Prelim_geo_pdf/SanPasqual_prelim.pdf

Hertlein, L.G., 1929, Stanford University Bulletin, 5th Ser., no. 78, p.82.

Hertlein, L.G., and Grant, U.S. IV, 1944, The Geology and paleontology of the marine Pliocene of San Diego, California, Vol. II, Pt. I, Geology: San Diego Society of Natural History.

Kennedy, G.L., 1973, Early Pleistocene invertebrate faunule from the Lindavista Formation, San Diego, California: San Diego Society of Natural History Transactions, v. 17, p. 119–128.

Kennedy, M.P., 1977, Geology of National City, Imperial Beach, and Otay Mesa quadrangles, southern San Diego metropolitan area, California: California Geological Survey Map Sheet 29, (1:24,000).

Kennedy, M.P., and Peterson, G.L., 1975, Geology of the San Diego metropolitan area, California: California Division of Mines and Geology Bulletin 200, 56 p., 6 plates (1:24,000).

Kennedy, M.P., and Tan, S.S., 2008, Geologic Map of the San Diego 30' x 60' quadrangle, California: California Geological Survey Regional Geologic Map Series, 1:100,000 Scale, Map No. 3 (1:100,000).

Kern, J.P., and Rockwell, T.K., 1992, Chronology and deformation of Quaternary marine shorelines, San Diego County, California: *in* Quaternary coasts of the United States: marine and lacustrine systems, SEPM Special Publication No. 48.

Kuper, H.T., and Gastil, G., 1977, Reconnaissance of marine sedimentary rocks of southwestern San Diego County, *in* Farrand, G.T., ed., Geology of southwestern San Diego County, California and northwestern Baja California: San Diego Association of Geologists Guidebook, p. 9–14.

Kyte, F.T., Gersonde, R., and Kuhn, G., 2005, Detailed results on analyses of deposits of the Eltanin impact, recovered in sediment cores from Polarstern expedition ANT-XVII/5A: 36th Annual Lunar and Planetary Science Conference, 36, 2129, 2p.

Kyte, F.T., Gersonde, R., and Kuhn, G., 2006, Sedimentation patterns of meteorite ejecta in Eltanin impact deposits at Site PS58/281: 37th Annual Lunar and Planetary Science Conference, 37, 2305, 2p.

Margolis, S.V., Claeys, P., and Kyte, F.T., 1991, Microtektites, Microkrystites, and spinels from a late Pliocene asteroid impact in the Southern Ocean: Science, v. 251, no. 5001, p. 1594–1597. http://we.vub.ac.be/~dglg/Web/Claeys/Pubs/Margolis-etal-91.pdf

Moore, G.W., 1972, Offshore extension of the Rose Canyon Fault, San Diego, California, in Geological Survey research 1972: U.S. Geological Survey Professional Paper, 800-C, p. C113–C116.

Peters, R., and Jaffe, B.E., 2010, Identification of tsunami deposits in the geologic record; developing criteria using recent tsunami deposits: U.S. Geologic Survey Open-File Report 2010–1239, 39 p. http://pubs.usgs.gov/of/2010/1239/

Peterson, G.L., 1970, Quaternary deformation of the San Diego area, southwestern California, in Pacific slope geology of northern Baja California and adjacent alta California: Pacific Sections of AAPG, SEPM, and SEG, p. 120–126.

Peng, H., 1994, An extraterrestrial event at the Tertiary-Quaternary boundary: KT event and other catastrophes: Lunar and Planetary Institute Contribution No. 825, p. 88–89.

Rhodes, B., Tuttle, M., Horton, B., Doner, L., Kelsey, H., Nelson, A., and Cisternas, M., 2006, Paleotsunami research: EOS Vol. 87, no. 21, 23 May 2006, p. 205–209.

Saul, R.B., and Wootton, T.M., 1983, Geology of the south half of the Mint Canyon Quadrangle, Los Angeles County, California: California Division of Mines and Geology, Open-File Report 83-24LA, 139 p.

Scheffers, A., and Kelletat, D., 2004, Bimodal tsunami deposits — a neglected feature in paleo-tsunami research: Geographie der Meere und Küsten, Coastline Reports 1 (2004) ISSN 0928-2734, p. 67–75.

Scheidemann, R.C. Jr., and Kuper, H.T., 1979, Stratigraphy and lithofacies of the Sweetwater and Rosarito Beach Formations, southwestern San Diego County, California and northwestern Baja California, Mexico: in Stuart, C.J. ed., A guidebook to Miocene lithofacies and depositional environments, coastal southwestern California and northwestern Baja California: Geological Society of America Guidebook for field trips Joint meeting with SEPM, Pacific Section, p. 107–118.

Squires, R.L., Groves, L.T., and Smith, J.T., 2006, New information on molluscan paleontology and depositional environments of the Upper Pliocene Pico Formation, Valencia area, Los Angeles County, southern California: Publication of the Natural History Museum of Los Angeles County, Contributions in Science no. 511,

Todd, V.R, Busch, L.L., Foster, B.D., Hernandez, J.L., and Tan, S.S., 2006, Geologic Map of the Ramona 7.5' Quadrangle, San Diego County, California: A Digital Database: California Geological Survey (1:24,000). ftp://ftp.consrv.ca.gov/pub/dmg/rgmp/Prelim_geo_pdf/Ramona_prelim.pdf

Varela, A.N., Richiano, S., and Poire, D.G., 2011, Tsunami vs Storm origin for shell bed deposits in a lagoon environment: An example from the upper Cretaceous of southern Patagonia, Argentina: Latin America Journal of Sedimentology and Basin Analysis, v. 18(1), p. 63–85.

Walsh, S. L., and Deméré, T. A., 1991, Age and stratigraphy of the Sweetwater and Otay formations, San Diego County, California, in Abbott, P. L., and May, J. A., editors, Eocene geologic history San Diego region: Pacific Section, Society of Economic Paleontologists and Mineralogists (SEPM), Pacific Section, Field Trip Guidebook, Book 68, p. 131-148.

Ward, Steven R., 2002, eltanin_ward.jpg:http://users.tpg.com.au/users/tps-seti/space-gd7.html#references *at* Australian Spaceguard Survey-website, http://users.tpg.com.au/users/tps-seti/eltanin_ward.jpg

Ward, S.R., and Asphaug, E., 2002, Impact tsunami-Eltanin: Deep-Sea Research Pt. II, 49, p 1073–1079.

White, D. R., 1983, Shoreface to lagoonal facies of the marine Pliocene lower Saugus Formation, Northern Simi Valley, California; in Squires, R. L. and Filewicz, M. V., eds., Cenozoic Geology of the Simi Valley area, southern California: Pacific Section Society of Economic Paleontologists and Mineralogists, Fall Field Trip Volume and Guidebook, p. 207–220.

Camp Pendleton and the San Onofre coast, January 22, 1990.
Photo: Woodrow L. Higdon, Geo-Tech Imagery.

PRELIMINARY ANALYSIS OF PIEDRA DE LUMBRE AND TALEGA CANYON CHERTS:
Distinctive and Historically Significant Outcrops on Camp Pendleton, San Diego County, California

Eleanora I. Robbins
Department of Geological Sciences
San Diego State University
San Diego, California 92182-1020

Andrew R. Pigniolo
Laguna Mountain Environmental, Inc.
7969 Engineer Road, Suite 208
San Diego, California 92111

Greg T. Cranham
Hargis + Associates, Inc.
9171 Towne Centre Drive, Suite 375
San Diego, California 92122

William J. Elliott
Consulting Engineering Geologist
P. O. Box 541
Solana Beach, California 92075

ABSTRACT

Two possibly related deposits of chert having prehistoric importance occur on Camp Pendleton. When knapped, these cherts hold a knife-sharp edge—a prized and valuable quality for early toolmakers. Tools and flakes made from this material are distributed in archaeological sites from the Hemet area in Riverside County to the international border with Mexico.

Porcelanitic sandstone of presumed Eocene age hosts the chert in Piedra de Lumbre and Talega Canyons. The chert is distinctive because of its clear angular quartz grains floating visibly in a microquartz matrix. Known outcrops occur as isolated ridge caps. The rocks sit on a paleosol of probable Paleocene age.

Petrographic details obtained from thin sections have been used to delineate the character of the Piedra de Lumbre and Talega Canyon chert and sandstone. In both, quartz grains are highly angular and unsorted; chalcedony occurs in vugs, cracks, and layers. Organic matter takes the form of microbial mats, linear arrays, filaments, algal spheres, and cysts. Accessory minerals are rare.

A hydrothermal spring origin for the deposits is implicated. Silica is not a common ion in solution, requiring heat, pH values greater than 8.5–9, or the presence of an easily dissolvable silica source such as a volcanic ash. Sources of silica in the vicinity include bentonite and kaolinitic clay. Lacking accessory minerals that are diagnostic of alkaline solutions, a hot spring origin is most plausible. The chert may have formed at the Fe^{2+} / Fe^{3+} redox boundary because hematite and pyrite co-exist in most samples.

Located at or near the subducting Farallon/North American plate margin, a source of heat does not appear to be problematic. The age of hot spring activity, however, is not well constrained. Volcanic rocks in the area include Tertiary andesite and dacite. Therefore, this region was not deficient in thermal sources to dissolve and carry silica to the sites of chert deposition.

INTRODUCTION

Two possibly related deposits of chert in porcelanitic, resistant siliceous sandstone have prehistoric importance and occur in the central and northern portions of the Camp Pendleton Marine Corps Base, northern San Diego County, California. The resistant units form cap rock on hillsides overlooking Piedra de Lumbre and Talega Canyons (Fig. 1). Although variable in hardness and appearance, the chert holds a knife-sharp edge when knapped, making it valuable to early toolmakers. Tools and flakes made from this material are distributed in archaeological sites from the Hemet area in Riverside County to the international border with Mexico, shedding light on trade routes and trading relationships among Archaic and Late Prehistoric people. In addition to its importance in the archaeological record, this chert is lithologically distinctive. It has large angular clear quartz grains that impart a luminous quality—a distinctive signature that aids in identifying its provenance at Camp Pendleton.

Informally designated Chert of Piedra de Lumbre and Chert of Talega Canyon, these outcrops are very limited in areal extent and thickness, and thus do not comprise geologically mappable units at the quadrangle scale. Although not formally named and delineated, the first known recognition of the Piedra de Lumbre outcrop by a geologist was by Furu (1982). The first description was provided by Cranham et al. (1994); at that time, the outcrop was thought to occur within the Pleasants Sandstone member of the Williams Formation (Yerkes et al., 1965). However, field examination by the authors at the Piedra de Lumbre locality suggests that these resistant rocks are more consistent with the basal unit of the middle Eocene Santiago Formation (Wilson, 1972; Weber, 1982).

Although the location of the source quarry was not known at that time, prehistoric tools made from this material were described as early as the 1970s (Pigniolo, 1994). Archaeologists identified the source location in Piedra de Lumbre Canyon several years later (Murray and Fenenga, 1981). A comprehensive study of the chert's distribution in the archaeological record was conducted by Pigniolo (1992), who identified seven individual exposures. The Talega Canyon deposit has only recently been recognized as being stratigraphically and petrologically similar to that in Piedra de Lumbre Canyon (Sikes et al., 2006). The generic term "Piedra de Lumbre Chert" is used as a descriptive, informal unit name, and is consistent with usage in the archaeological literature.

Several explanations for the genesis of chert-bearing rocks at Piedra de Lumbre Canyon have been suggested, including volcanic (Furu, 1982), an ignimbrite deposit (Cranham, 1985), or a hot springs deposit (Dr. Mike Walawender, personal commun., 1994). Because geologic aspects of the deposit have been little studied, consensus was lacking. The objective of this study is to propose a genetic interpretation of the Piedra de Lumbre and Talega Canyon cherts based on a comprehensive review of available data, reexamination of outcrops and thin sections, and comparison with similar outcrops having a known mode of formation. In addition, the importance of these outcrops in the archaeological record is summarized in the next section.

The field area is bounded by the Elsinore and Newport-Inglewood-Rose Canyon Fault Zones (Cranham et al., 1994; Jennings, 1994). Nearby Tertiary volcanic rocks include dacitic necks at Morro Hill and

Cerro de la Calavera (Oligocene?), and an andesite neck near Horno Summit (Miocene?) (Larsen, 1948; Weber, 1963; Cranham et al., 1994; Dr. Tom Deméré, oral commun., 2012). Ash in the form of bentonite or glass shards has been noted in Eocene, Oligocene, and Miocene rocks and sediments of San Diego County (Elliott and Berry, 1991; Berry, 1999; Cranham et al., 1994); as many as three tuff beds have been mapped in the upper member of the Santiago Formation on Camp Pendleton (Cranham et al., 1994). Digital geologic maps that cover the areas discussed herein have been published online by the California Geological Survey for the Las Pulgas Canyon, Morro Hill, San Clemente, San Luis Rey 7.5' quadrangles.

ARCHAEOLOGY

While Monterey and Franciscan Formation cherts served as dominant sources of lithic material for prehistoric production of flaked stone tools from Orange County through the central coast region, sources of chert in western San Diego County are generally limited to small outcrops of limited quality material or sparse secondary deposits. Both the Piedra de Lumbre (CA-SDI-10008/10708) and Talega Canyon (CA-SDI-13655) outcrops are archaeological sites where chert was used as flaked lithic tool material by prehistoric Native Americans. Lithic material at Piedra de Lumbre Canyon was of sufficient quality and quantity to make it one of only three regionally important chert sources in San Diego and Imperial Counties, along with the Rainbow Rock and Cerro Pinto Wonderstone sources in the western Colorado Desert. The coarse-texture, which reflects the sandstone protolith of the Talega chert, created less than ideal conchoidal fracture. This, combined with the more limited amount of material at the source, reduced distribution of Talega chert to a more local level, closer to the seasonal mobility range of directly proximate groups.

Artifact Types and Age

Early dart point styles of Piedra de Lumbre chert at the Windsong Shores archaeological site in Carlsbad (Gallegos and Carrico, 1984) indicated that Piedra de Lumbre chert was used by Archaic people at least 8,390 years before present. The presence of tools called cresentics made from Piedra de Lumbre chert elsewhere suggests even earlier use.

Although larger bedrock material is obtainable, fracturing at both the Piedra de Lumbre and Talega outcrops limits average high quality chert fragments to roughly 10 cm in diameter. Average flaked artifacts of all lithic materials at archaeological sites with Piedra de Lumbre chert was 7.57 grams, while the average weight of Piedra de Lumbre chert alone was only 1.95 grams (Pigniolo, 1992). Limited rock size had an effect on types of stone tools that could be produced from the material. The vast majority of prehistoric stone tools produced from Piedra de Lumbre chert are projectile points and knives followed by retouched flake tools. The geographic name Piedra de Lumbre (Stone of Light or Illumination) may have either been inspired by the luminous chert itself or by historic use of the material for gunflints (Pigniolo, 1994).

Prehistoric technology changed through time, and with the introduction of bow and arrow technology in the Late Prehistoric period, projectile points were smaller in size and generally less than 3.5 grams in weight. The reduced projectile point size allowed Piedra de Lumbre chert to be a more useful tool material, so the intensity of Piedra de Lumbre chert use increased dramatically. During the Late Prehistoric, production and exchange of the material became more widespread. Several larger Late Prehistoric sites in the Camp Pendleton area have as much as 60 to 70 percent Piedra de Lumbre chert flaked stone material (York and Wahoff, 2009).

Distribution

Piedra de Lumbre chert is distributed in archaeological sites from the Hemet area in western Riverside County to the Irvine area in Orange County and south to near the International Border. Ease of conchoidal fracture as well as the usual ochre color and characteristic sparkling of the enclosed clear angular quartz grains may have added to the attraction. Only obsidian has a wider distribution in the coastal region. While a variety of factors affect the abundance of Piedra de Lumbre chert at particular archaeological sites, the Late Prehistoric distribution appears to reflect a fall-off pattern, which supports energy-distance models of decreasing abundance with distance from the source (Pigniolo, 1992). By the end of Late Prehistoric time, cultural boundaries acted as barriers to direct procurement of material from the sources.

GEOLOGIC SETTING

The Piedra de Lumbre and Talega Canyon localities were examined by the authors in April and June 2012. Access to Piedra de Lumbre outcrops was limited; the eastern slopes of the ridge with outcrops falls within an active ammunition storage bunker area. Delineation of the Santiago Formation in this area (Fig. 2 and Cranham et al., 1994) should therefore be considered approximate. The Talega Canyon outcrop(s), which consists of three boulders (Sikes et al., 2006), was inaccessible due to the presence of an active live-fire practice range near the ridge-top site. Although the actual outcrop could not be observed directly, float was collected along the north side of the ridgeline and from the dry creek below.

Outcrop Geomorphology

At the Piedra de Lumbre locality, in-place resistant beds and/or displaced erosion-resistant float occur in nine separate areas. The authors had access to all sites except distant area 9 (Fig. 2). These prominent outcrops occur as scattered, discontinuous patches along the crest of a northerly-trending ridgeline (Figs. 2 and 3).

The contiguous outcrop at areas 1 through 3 was the only site investigated in detail. There, nodular chert is multicolored (gray, olive brown, yellowish, reddish, and rarely black); it is extremely resistant, forming sharp contacts with the host porcelanitic sandstone. Chert also occurs as extensive scatter at the exposed surface of the porcelanitic sandstone.

Overlying strata have been largely removed by erosion, leaving a stripped dip slope or cuesta. The resistant strata dip approximately 10° in a generally westerly direction (Fig. 4) and do not extend far down the western slope, having been been removed by erosion or landsliding. Mass wasting, including landsliding has moved scattered erosion-resistant remnants down the relatively gentle western topographic slope.

An approximately 5x20 m, intact block of Piedra de Lumbre sandstone was observed well down the western slope of the asymmetrical ridgeline from areas 1 through 3; this block was identified as a landslide by its dislocation from the main outcrop, lower elevation, and approximately 5° backward rotation in a northeasterly direction (Fig. 4). Other landslide blocks and disconnected clusters of float were similarly identified on the western side of the ridgeline (e.g., areas 4, 7, and 8 as shown in Fig.2).

The steeper eastern slope of the ridge exposes underlying strata. Much of the eastern slope of the ridge is composed of Upper Cretaceous Williams Formation (Yerkes et al., 1965), dominated by reddish to grayish brown arkosic to sublitharenite sandstone interbedded with siltstone and mudstone, commonly exhibiting graded bedding (Cranham et al., 1994).

Lithologic Descriptions

The uppermost resistant unit and immediately underlying strata are described here from exposed cross-sectional edges on the western side of the ridge top. Based on field observations, the outcrop can be subdivided into four distinct units (Fig. 5,1). The uppermost resistant rock, forming the ridge cap, is designated "Unit 1." From upper to lower, Units 1 through 4 are described as follows:

Surface: Discontinuous layer of fractured chert scattered on resistant units and along ridge top. Whether this layer of chert is in situ but fractured, is residuum from weathering, or is scattered because of human use could not be assessed.

Unit 1: Massive bedded, cream-white porcelanitic sandstone, with matrix supported fine- to medium-grained quartz grains, forms a tabular cap along the ridgeline. Thickness is approximately 0.76 m, ±0.15 m. The unit is locally stained red, orange, or yellow. Porcelanitic sandstone has a conchoidal fracture, and is very hard, massive, and nonporous. Fresh hand samples break across quartz grains. Dark minerals are rare. No textural or sedimentological features were observed in hand samples that would assist with determining genesis. The lower part of the unit is less siliceous, and more chalky, probably the result of a kaolinite-rich matrix. The lower contact varies from sharp and undulatory to indistinct and gradational.

Chert in the porcelanitic sandstone (Fig. 5,2) occurs as olive-brown nodules as large as 25 by 40 cm. None has been observed in three-dimensions. Contacts with the porcelanitic sandstone are clean and often bounded by fractures.

Porcelanitic sandstone and olive-brown chert collected in Talega Canyon appear similar in hand specimen to "Unit 1" at the Piedra de Lumbre locality.

Unit 2: Iron-stained white sandstone. Maximum thickness is approximately 15 cm. Unit is fractured and cavernous. Iron staining is intense, in the manner of hydrothermal alteration. Unit 2 is the upper weathered portion of "Unit 3", described below. Lower contact is highly irregular.

Unit 3: Massive bedded, medium- to coarse-grained quartz sandstone. Freshly broken surfaces vary in color from dull cream white to earthy chocolate brown. Weathered surfaces are typically light tan to milky white. Fresh specimens break around quartz grains, which are generally sub-rounded to rounded. Quartz grains are both grain- and matrix-supported. Matrix appears to be dominated by clay, probably weathered from feldspar. Pore space is visible, giving the rock an estimated porosity of 10%. Thickness is approximately 1.5 to 1.8 m.

This unit is consistent with the informally defined lower member of the middle Eocene Santiago Formation (Wilson, 1972; Weber 1982) mapped in hills west of the study area (Cranham et al., 1994) (Figs. 2 and 3). These typical exposures of Santiago Formation sandstone are composed of quartz, feldspars, and typical accessory minerals that are so common in sub-tidal estuarine sand-bar environments (Abbott, 1985). In contrast, the Piedra de Lumbre sandstone suite appears to be highly weathered and/or hydrothermally altered, and lacks detrital feldspar.

Unit 4 (base): Highly weathered, brown to yellowish-brown, massive- to thinly-bedded, clayey and silty sandstone, composed of sub-rounded to rounded, detrital quartz grains. Bedding appears to be conformable with overlying unit.

Unit 4 is correlated with the Pleasants Sandstone member of the upper Cretaceous Williams Formation (Yerkes and others, 1965; Cranham et al., 1994). Regionally, total thickness exceeds 427 meters. Most of the feldspar, so common in other Williams Formation sandstones, appears to have weathered to clay. Furthermore, the few thin inter-beds of lavender to bluish gray silty to clayey shale may have acted as slip surface(s) allowing overlying units to slide downslope to the west (Fig. 4).

Deep weathering of the uppermost part of the Williams Formation has been observed elsewhere in the region, where a distinctive kaolinite-bearing paleosol (Abbott, 1999) developed prior to deposition of the Santiago Formation (Cranham et al., 1994). Intense weathering (and/or hydrothermal alteration) of the lower Santiago Formation, as observed in Units 2 and 3, has not been observed elsewhere at Camp Pendleton. It likely represents a poorly understood, localized phenomena.

Sedimentary Structures

Using data from modern hot springs as proxy reference images, the authors searched for megascopic sedimentary structures in the field. For example, modern hot springs are characterized by vent geyserite, botryoidal structures, stromatolite heads, and laminations where silica-charged water flows into quiescent pools (Hinman and Walter, 2005). Laminations were the only megascopic siliceous sedimentary structures observed; these laminated chert samples were restricted to Area 6 of the Piedra de Lumbre site (Fig. 5,3).

PETROGRAPHY

Materials and Methods

Rocks assessed in the petrographic analysis were collected by archaeologists to determine provenance of source material at archaeological sites (Table 1). Nineteen chert artifacts and rocks, along with two sandstones were thin sectioned for those analyses.

Megascopic Observations

Megascopic observations from thin sections provide important information (Table 1). Brown and orange brown are the dominant colors. Brown is generally imparted by dispersed organic matter, while orange brown is from iron-oxide minerals. Colorless rhythmic laminations and brown-outlined convoluted laminations form megascopic sedimentary structures. Opaque minerals are so abundant in some thin sections that they can be observed in patches and arrays, along with organic-rich concentrations.

Dominant Mineralogy. Table 2

Quartz grains provide the framework and microquartz is the matrix in most of the analyzed samples (Fig. 6,1). Four samples are classified as quartz wackes, four are quartz arenites, and 13 are chert arenites. In all, quartz grains are unsorted, with shapes typically ranging from angular to subangular; a small percentage were found to be subrounded to rounded.

Accessory minerals are rare, comprising less than 1% of the assemblages. Zircon (some euhedral), hornblende, and pyroxene are present in more than half the samples. Other minerals are rarer still.

Lithic fragments include polycrystalline quartz, undulatory quartz, and mylonitic quartz (Fig. 6,2). "Ash" (Fig. 6,3) is present in half the thin sections, but is also rare.

The quartz wacke is cemented with some presently unidentified clay. The rock itself is white, suggesting that the clay may be kaolinite.

Interpretation: The angularity of the quartz and the mineralogy of the accessory minerals suggest that the provenance was close. Composition of the accessory minerals suggests a volcanic source.

Microquartz. Table 2, Table 3.

The matrix of most samples is microquartz (Fig. 6,4). Microquartz crystallites range from 2-5 µm. The matrix is cut by cracks (Fig. 6,5), veins (Fig. 6,6), and vugs (Fig. 6,7) that are also filled with microquartz, along with chalcedony (banded or spherulitic) and authigenic quartz in the form known as dogtooth (Fig. 6,8).

Interpretation: Cracks crossing other cracks (Fig. 6,5), cracks cutting vugs (Fig. 6,9), and vugs having multiple layers of chalcedony and dogtooth quartz (Fig. 6,8) are all evidence for multiple silica injection (Dr. Gary Girty, written commun., 2012).

Organic Matter Types and Distribution. Table 4.

Organic matter is abundant in 28% of chert arenite thin sections, and rare to lacking in quartz wacke sections. The organic matter is dominantly in the form of microbial mats (Fig. 7,1-3), single laminae (Fig. 7,4), filaments of cf. *Phormidium* sp. (Fig. 7,5) (a cyanobacterium of hot springs), algal filaments (Fig. 7,6), unidentified algal- and fungal-sized spores (Fig. 7,7) and algal? cysts (Fig. 7,8). Pellets resembling zooplankton fecal pellets (Fig. 7,9) are also present.

Interpretation: Microbial mats in springs are typically formed by photosynthetic cyanobacteria. Their preponderance supports an aqueous environment for deposition. High temperature hot springs such as at Yellowstone National Park are also colonized by chemolithotrophic hydrogen oxidizers (Blank et al., 2002) that are filamentous and would be expected to form single laminae.

Pyrite Types and Distribution. Table 5.

Pyrite enmeshed in the organic matter is in the form of 1 µm octahedrons (Fig. 7,1), typical of sulfate-reducing bacteria that use organic carbon as their C source (Folk, 2005). Framboids (Fig. 7,10), another pyrite type precipitated by bacteria (Kyle and Schroeder, 2005; Maclean et al., 2008), occurred in more than 50% of the thin sections.

Interpretation: Two generations of pyrite precipitation can be defended. The first is pyrite enmeshed in organic tissues. Secondly, pyrite in veins and vugs suggest an additional period of anoxia.

Iron Oxide Mineralogy and Distribution. Table 5.

In thin sections, hematite and amorphous phases of iron oxides are observed as linear, localized, and interstitial configurations, as well as crack fillings (Fig. 7,11), attesting to a complex history. Fully crystalline hematite is also present. Microbial minerals in the form of iron bacteria (Robbins and Norden, 1994; Robbins and Hayes, 1996) are present: cf. *Leptothrix discophora* holdfasts, cf. *Siderocaspa* spheres and capsules (Fig. 7,12), cf. *Siderocystis* agglomerations along filaments.

Interpretation: Hematite and iron oxides in cracks that cut across silica-filled cracks and vugs suggest a later input of iron (Fig. 6,10).

Geopetal Fillings

A characteristic of hot spring deposits is a form of glass that fills cavities (Hinman and Walter, 2005). These geopetal fillings (Fig. 6,11) (Table 2) were abundant in one quartz wacke from the Piedra de Lumbre site; as many as 10 individual episodes of filling can be counted in one cavity. In another quartz wacke, deformation has created elongate forms that resemble geopetal fillings.

Flow Structures

Convoluted laminations are present in four chert arenite samples (Table 1). They are outlined by a thin layer of organic matter and are filled with microquartz. Another possible flow structure in PDL 11 (Pigniolo, 1992, 1994) has elongated microquartz blebs slightly oriented in one direction (Fig. 6,1).

Interpretation: Convoluted laminations are typically formed when microbial mats are ripped out and transported downstream from their point of origin.

Inorganic Laminations and Structures

Rhythmic bedding characterizes two chert thin sections (Table 1). The draping of one band of inorganic laminations in PDL 7 allows determination of orientation (Fig. 6,12). The descending sequence of silica in one band is: microquartz, chalcedony, dogtooth quartz; in another it is dogtooth quartz, chalcedony, dogtooth quartz (Fig. 6,13). In a second chert (PDL 3), the laminations are less obvious and the orientation equivocal; the silica sequence is microquartz and dogtooth quartz (Fig. 6,10). Siliceous pisolites are also present (Fig. 6,14).

Interpretation: Very quiet water deposition is implicated for inorganic laminations. Rhythmic bedding could be seasonal and result from deposition during dry seasons.

CONCLUSIONS

Interpretation of Deposit

Geological model. The shoreline of the tectonically active Pacific margin is characterized by numerous episodes of sea level changes (Abbott, 1999). Following subaqueous deposition of the Upper Cretaceous Williams Formation (Yerkes, 1965), either sea level fell or uplift occurred allowing for subaerial formation of a paleosol during the Paleocene (Abbott, 1999). Then, with sea level rise in the Eocene, clean, near-shore sheets of quartz sand (Unit 3) were deposited conformably over the paleosol which had been previously weathered into the upper meter or so of the Williams Formation.

Subaerial exposure followed, which allowed atmospheric weathering to create a pedogenic horizon with irregularly undulating upper and lower surfaces. This poorly defined horizon is herein named Unit 2. With Units 2 and 3 in place, and all beds still horizontal and subaerially exposed by falling sea level and/or isostatic uplift, hydrothermal activity along an undefined rift in the earth's surface created an approximately north-south line of isolated pools or small lakes fed by hydrothermal springs into which Unit 1 was deposited.

Being at or near the Farallon/North American plate margin (Atwater, 1970), it is conjectured that some intermittent explosive volcanic activity rained angular clear quartz randomly into the thermal pools, thus accounting for at least some of the dispersed and matrix-supported grains in the white porcelanitic sandstone. Hydrothermal activity continued sporadically for some unknown time, accounting for the several episodes of silica emplacement and formation of chert nodules.

Then, either sea level rose and/or the land sunk enough to allow continued deposition of Santiago Formation conformably over the isolated ponds/lakes. Eventually, the lower and middle Miocene San Onofre Breccia (Stuart, 1975) was deposited conformably over the Santiago Formation. These strata were then uplifted and tilted gently westward. Finally, mass wasting and erosion during Quaternary time exposed the rocks as seen today at Camp Pendleton.

Hot spring origin model. A hot spring setting is the likely origin of the chert deposits. Hot spring deposition could account for authigenic microquartz in the matrix as well as in vugs, cracks, and veins. The presence of microquartz laminations, microbial mats, single laminae mats, pisolites, and geopetal void fillings all support a hot spring origin, even though no macroscopic botryoidal, stromatolitic sedimentary structures, or vent openings were observed in outcrop.

Microquartz laminations, as well as microbial mats of photosynthetic cyanobacteria, such as *Phormidium* sp., are well known from modern hot springs (Hinman and Walter, 2005). Depending on temperature, microbial populations using hydrogen as their energy source are common in modern hot springs (Blank et al., 2002) and could be expected to proliferate in subsurface realms as long as soluble sources of carbon were also present.

Sources of silica. Several potential sources of easily dissolved silica occur in the area, including volcanic ash, bentonite, and kaolinite. Silica in solution requires heat or pH values greater than 8.5–9 (Hem, 1989), thereby providing useful analog data that could not have been determined by petrology alone.

Sources of heat. Varying amounts of volcanic activity along the west coast of North America occurred during early- mid-Tertiary time (Jennings, 1977; and other map sheets in that CDMG Series). Horno Summit, Morro Hill, and Cerro de la Calavera are volcanic outcrops that lie within about 20 km of Piedra de Lumbre, and within about 40 km of Talega Canyon. These potential volcanic sources of heat have been dated as Eocene through Miocene (Elliott and Berry, 1991; Berry, 1999; Cranham and others, 1994).

Age of deposit. While no age diagnostic fossils have been found associated with the chert or underlying sandstone (Units 1 through 3), the age of these deposits can be bracketed stratigraphically. They are clearly younger than the underlying Upper Cretaceous Pleasants member of the Williams Formation (Yerkes and others, 1965), as well as the underlying Paleocene paleosol (Abbott, 1999). They are older than the overlying middle Eocene Santiago Formation (Wilson, 1972; Weber, 1982).

Episodes of early to middle Tertiary volcanism and elevated heat flow, necessary to support a hot spring genesis for Unit 1, lend further support for an early Tertiary age.

Generations of silica input. Several generations of silica input are represented by the microquartz. *Phase 1*: Silica cement that forms microquartz between angular quartz grains was the initial phase. The quartz grains are not in mutual contact, suggesting that this phase of silicification occurred as soft sediment replacement before lithification of the host sandstone; something like a slurry is envisioned. There were, however, no observed ghosts of siliceous organisms, such as diatoms, to support this thesis that silicification was the result of aqueous solutions. *Phase 2*: The next silicification phase was through vugs that represent loci of point source discharge; vugs are lined with microquartz, chalcedony, and authigenic dogtooth-type quartz. *Phase 3*: The final silicification phase involved solutions carried through cracks and veins that cut across both vugs and microquartz matrix. These cracks and veins are filled with microquartz, chalcedony, and authigenic dogtooth-type quartz.

Silicification timing. Insight obtained from study of fossil hot springs elsewhere suggests that timing phases for silicification may vary from contemporaneous to annual. At Steamboat Springs, Nevada, Hinman and Walter (2005) showed that dehydration during the dry season induced precipitation of microquartz precursors. Analysis of hot springs at Yellowstone showed that silica deposition coincides with drop in temperature (Guidry and Chafetz, 2002).

Iron dynamics. Reduced iron in the form of pyrite provides information about anoxic areas in the former hot springs, especially massive pyrite. In contrast, pyrite octahedrons and framboids that are enmeshed in the organic tissues attest to local anoxia. Furthermore, the coexistence of reduced pyrite and oxidized iron oxide phases in 20 of the thin sections suggests that the iron redox boundary was close to the surface during microquartz deposition. Distribution of iron oxide embedded in microquartz cement suggests that iron was carried by the silica-bearing solutions. Furthermore, the presence of iron oxide with no authigenic silica in cracks and veins suggest that iron was introduced again at a later time by circulation of post-depositional ground water.

Differences and similarities between Piedra de Lumbre and Talega Canyon. In contrast to Piedra de Lumbre cherts, Talega Canyon cherts are generally darker in color, have larger quartz framework grains, and exhibit bimodal microquartz. Similarities at both localities include: organic matter content, iron-oxide mineralization, and the presence of cracks, veins, and vugs filled with similar silica phases. Quartz grains at Talega Canyon are slightly larger, suggesting it may have been closer to its source of quartz.

Former subaerial distribution. It is likely that the white sandstone was deposited over a much larger area. The only remaining outcrops, however, currently form resistant outcrops on ridges where the sandstone was cemented with microquartz. It is possible that each surviving outcrop hosted its own hot spring. Similarly, modern hot springs are not solitary occurrences (Guidry and Chafetz, 2002), but often occur in widely separated groups or clusters. Scattered chert float occurs on outcrop surfaces, along hillsides, and in water courses, indicating that chert is the most resistant remnant of previously more extensive sandstone beds.

Surficial incorporation of quartz into bedded chert. The process(es) by which the Piedre de Lumbre and Talega Canyon silica solutions incorporated quartz grains are difficult to discern. A top-down surficial process is easiest to envision: a hot spring that discharged silica-bearing solutions along the surface of a volcaniclastic sandstone could have precipitated chert. This process might account for the chert at the surface of the porcelanitic sandstone, as well as organic matter in the chert that would be the remains of photic zone organisms. Additional quartz grains could have been carried to the site of deposition by being blown in by the wind, blown in from volcanic activity, carried by seasonal surface water flow, and/or plucked from the underlying unit.

Subsurface incorporation of quartz into nodular chert. The subsurface process of incorporating quartz grains into nodular chert has no simple or obvious solution. The fact that the chert carries visible clear angular quartz grains is what makes the Piedra de Lumbre and Talega Canyon cherts so distinctive; it is a very unusual type of workable material. Replacement of unconsolidated volcaniclastic sandstone during diagenesis is a possibility, and one can envision gravity-driven physical diffusion down through the sediments, creating a silica front and an accompanying porcelanitic luster. Replacement in the pore spaces would be a thermo-chemical process with reactions occurring between microquartz and clay. Concentrations of chert nodules in the subsurface cavities would occur where fractures and favorable flow paths had become established in the sandstone. Such piping structures would have had to be wide enough for heated water to carry quartz grains.

Subsurface incorporation of organic matter into nodular chert. The next problem, equally as complex, is explaining how the nodular chert acquired organic matter such as microbial mats. Because of the presence of organic matter, chert nodules cannot be simple replacement of porcelanitic sandstone. Microbial mats that are formed by photosynthetic organisms requiring access to sunlight would have no simple pathway into chert nodules. However, organisms using an alternative energy source such as hydrogen are common in modern hot springs. It seems reasonable to suspect that at least some of the organic matter contained within chert is remains of chemolithotrophic organisms that lived in pore spaces, where they could have remained catalytically/ biologically active as long as water and energy sources were present. Such microorganisms need carbon sources, which had to have migrated downward with silica-bearing fluids from above. Gravity-driven physical diffusion of hot spring solutions, carrying silica, quartz grains, and organic matter down through undefined conduits in sediments could provide at least one mechanism for observed chert nodules along apparent crevices and fractures. This could explain nodules of chert having connection to the surface, but not those lacking connections.

Alternative Explanations

The authors also considered and rejected other depositional environments that are known to deposit silica:

Marine: biosiliceous sediments. Evidence against this interpretation includes: patchy distribution of the host sandstone, presence of cyanobacterial filaments, microbial mats, lack of ghost structures of planktonic siliceous organisms, and lack of megascopic marine fossils. Evidence for is: porcelanitic sandstone with chert nodules, a characteristic of the marine Monterey Formation (Behl, 2011).

Volcanic: pyroclastic tuff. Evidence against this interpretation includes: monomineralic, no evidence of widespread distribution, no lithic fragments of rocks, no vitric shards or remnant ghosts of shards from devitrification; finally, the preponderance of angular quartz and the lack of tuffaceous lithic fragments and glass shards is inconsistent with a pyroclastic origin (Dr. Vic Camp, written commun., 2012). Evidence for is: potential hydrothermal alteration of underlying unit and angularity of the quartz.

Pedogenic: silcrete (soil silicification). Evidence against this interpretation includes: units coherent and tabular, no evidence of pedogenic rinds around grains, no bridging structures between quartz grains (Wopfner, 1983; Soil Survey Staff, 1999; Watts, 2006). Evidence for is: porcelanite has been reported in Australian silcrete.

Sedimentary: silcrete (playa lake). Evidence against this interpretation includes: rhythmic laminations only in one area of Piedra de Lumbre (Area 6), and no evidence of grain settling. Evidence for includes: clay in subsurface would serve as an aquitard, presence of minerals produced by iron oxidizing and sulfate reducing bacteria suggesting deposition at a redox boundary.

Meteor impact breccia: late stage ash-fall. Evidence against this interpretation includes: no shocked quartz (however, unit was sampled for archaeological information rather than stratigraphic data which would have been sampled in a systematic manner), no rock fragments, and no evidence for widespread distribution. Evidence for is: two large circular structures offshore (Legg et al., 2004) share similarities with other impact structures.

Future research needed

Information still needed to fully explain this deposit includes identification of clay minerals, thin sectioning of rocks selected for stratigraphic information and discerning origin of clear angular quartz grains floating in porcelanitic sandstone. More field work is needed to assess other areas of Camp Pendleton, particularly ones having similar exposures of ridgeline cap rock. Finally, the genetic relationship of nodular chert, organic matter, and abundant angular quartz grains remains a puzzle.

DEFINITIONS

Archaeology:
   Archaic: 8,400 to 1,300 years BP (6450 BC to 650 AD).
   Late Prehistoric: 650 to 1769 AD.

Silica:
   "*Ash*": quartz grains having embayments or holes.
   AQ, *Authigenic quartz*: some call this macroquartz or dogtooth; euhedral crystals 2–10 µm; fills vugs and cracks; rAQ, random orientation of authigenic quartz; pAQ, authigenic quartz perpendicular to crack or vug wall.
   CQ, *Chalcedony*: cryptocrystalline quartz, fibrous; banded or spherulitic.
   Chert: in hand specimen, dense, hard, siliceous rock, semivitreous to waxy luster, conchoidal fracture; holds a good cutting edge.
   MQ, *Microquartz*: some call this chert or microcrystalline quartz; authigenic, <10 µm; equant and anhedral; forms the matrix of most of the thin sections; also fills vugs and cracks.
   Mylonitic quartz: grains composed of stretched out interlocking quartz grains.
   Pisolites: siliceous spherules.
   PQ, *Polycrystalline quartz*: grains composed of interlocking quartz grains that go into extinction separately.
   Porcelanite: dense siliceous rock having texture, dull luster, hardness, conchoidal fracture, and general appearance of unglazed porcelain; less vitreous than chert.
   SQ, *Strained quartz*: quartz grains having undulatory extinction.
   TQ, *Transported quartz*: quartz grains in the chert arenite, quartz arenite, and quartz wacke.

Iron:
   Hematite: bright red in reflected light, hexagonal in shape.
   Iron oxide: orange in reflected light, amorphous, typically ferrihydrite.
   Pyrite: brassy yellow in reflected light; opaque in transmitted light; in forms of octahedrons (1 µm in size), cubes, or framboids (10-20 µm, composed of 1 µm octahedrons), or massive varieties.

ACKNOWLEDGEMENTS

The authors would like to thank the following for assistance with this project:

Dr. Rick Behl (California State University Long Beach), Dr. Vic Camp (San Diego State University, SDSU), Dr. Tom Demére (San Diego Natural History Museum), Dr. Becky Dorsey (University of Oregon), Alex Easton (Graphics Dept., Hargis + Associates), Bob and Suzanne Emery (Kumeyaay Ipay Interpretive Center), Dr. Gary Girty (SDSU), Nancy Hinman (University of Montana), Gwen Kenney (Environmental Security, Camp Pendleton Marine Corps Base), Dr. Dave Kimbrough (SDSU), Dr. Angela Lang (U.S. Geological Survey), Ad Muniz (San Diego Archaeological Center), Danielle Page (Environmental Security, Camp Pendleton Marine Corps Base), Dr. Aaron Pietruszka (SDSU), Dr. Kevin Robinson (SDSU), Dr. Tom Rockwell (SDSU), David Smith (retired), and George Zach (SDSU). We wish to especially recognize the late Dr. Michael J. Walawender, who provided guidance to authors Cranham and Pigniolo as part of their respective thesis projects, particularly with respect to the petrogenesis of the Piedra de Lumbre outcrop.

ILLUSTRATIONS

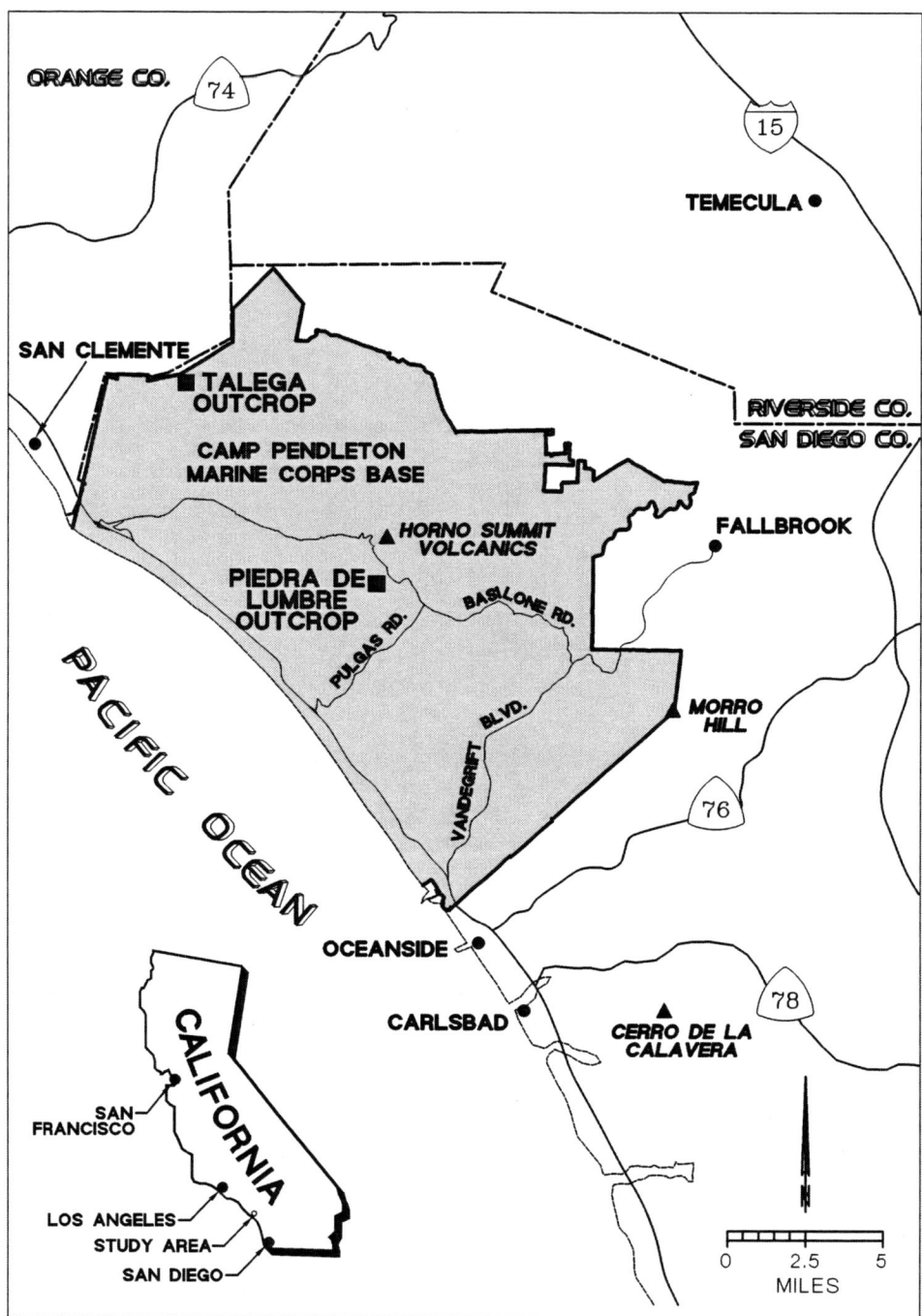

Figure 1. *Regional Location Map.*

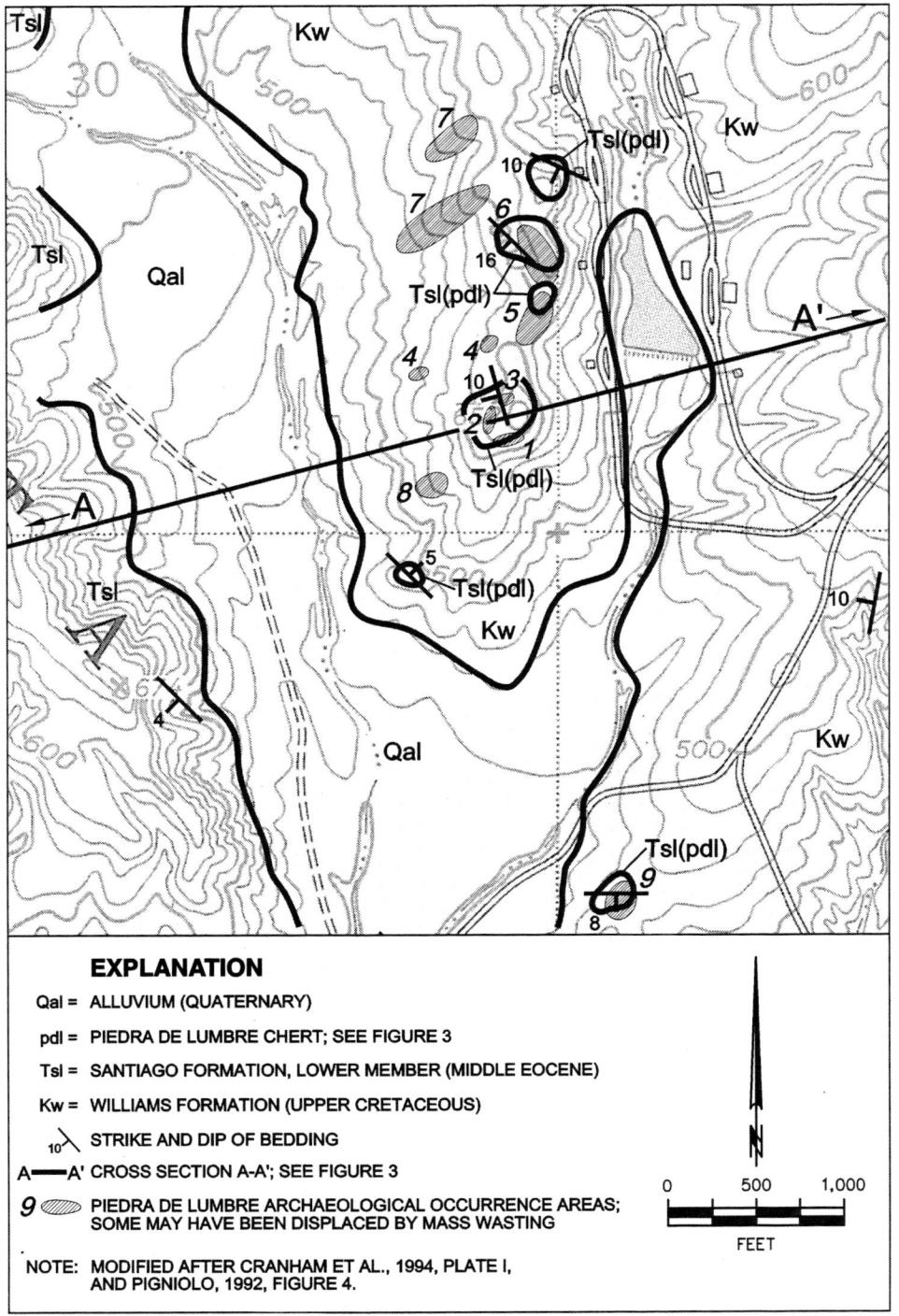

Figure 2. *Geologic Map, Piedra de Lumbre Outcrop Areas.*

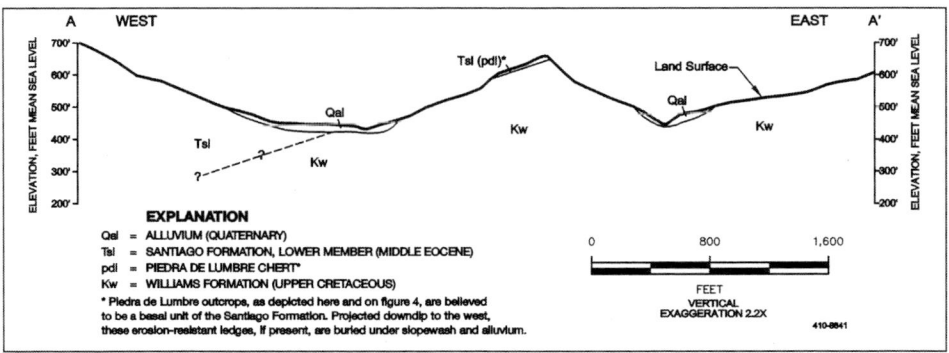

Figure 3. *Cross section A-A'.*

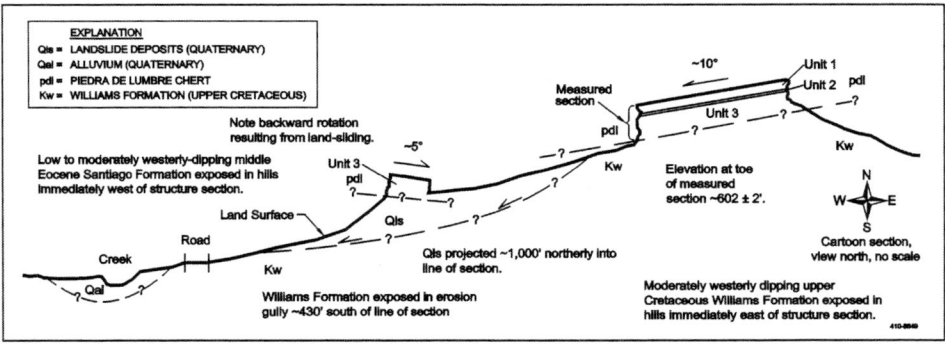

Figure 4. *Conceptual Cross Section.*

Figure 5. *Outcrop and rock photographs. 1. Outcrop at Area 1. 2. Hand specimen showing contact between porcelanite and chert. 3. Outcrop at Area 6 behind fence.*

PRELIMINARY ANALYSIS OF PIEDRA DE LUMBRE AND TALEGA CANYON CHERTS

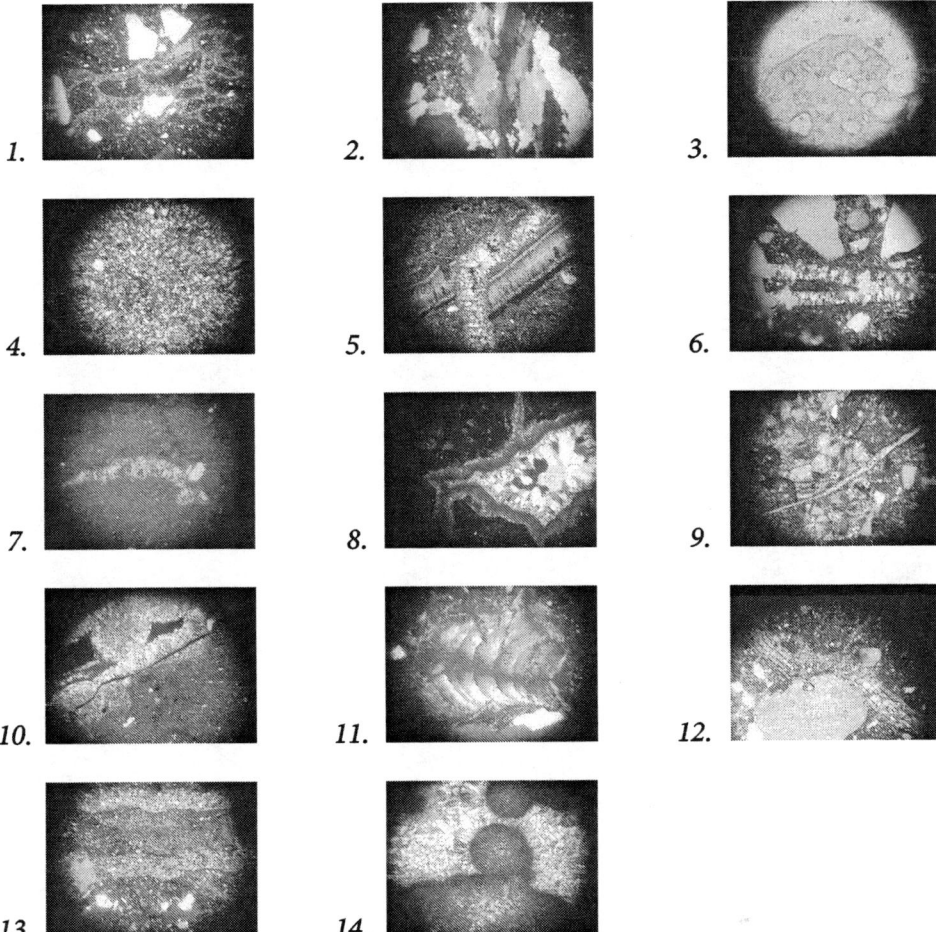

Figure 6. *Photomicrographs of siliceous minerals and structures (cross polarized light).* **1.** *PDL 11 Rounded microquartz structures elongated in a flow direction; white grains angular and sub-angular quartz (40×);* **2.** *Talega 4 Quartz mylonite grain (400×);* **3.** *Talega 2 Quartz grain dissolving in the manner of "ash" (400×);* **4.** *PDL 2 Microquartz (100×);* **5.** *PDL 3 Crack filled with chalcedony offsetting empty crack (40×);* **6.** *Talega 2 Crack filled with "dogtooth" quartz splitting into veins (40×);* **7.** *PDL 7 Vug filled with microquartz and perpendicular authigenic quartz (40×);* **8.** *Talega 7 Vug filled, from outside in, with microquartz, chalcedony, perpendicular authigenic quartz, and "dogtooth" quartz in center (40×);* **9.** *Talega 1 Microquartz-filled crack cutting vug (100×);* **10.** *PDL 10 Crack filled with iron oxide cutting vug (40×);* **11.** *PDL 6 Geopetal infilling of glass (100×);* **12.** *PDL 7 Complex siliceous lamina draping over "ash" clast (40×);* **13.** *PDL 7 Siliceous rhythmic laminations (40×);* **14.** *Talega 7 Siliceous pisolites (100×). See note on p. 113.*

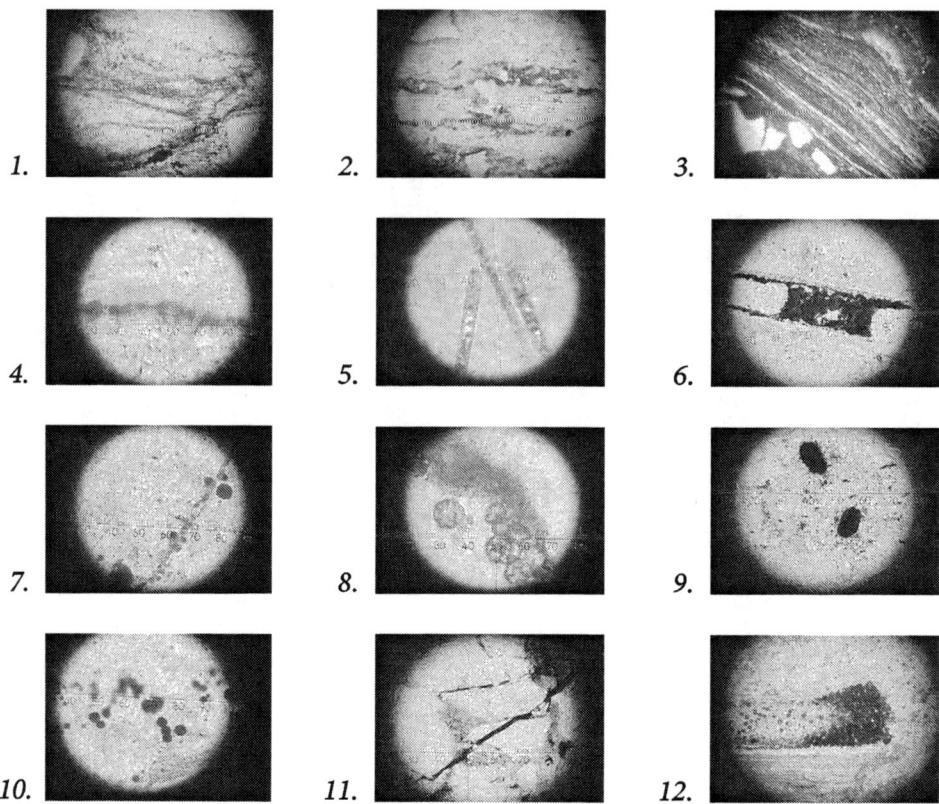

Figure 7. *Photomicrographs of organic matter and iron precipitates (all plane polarized light unless otherwise noted).* **1.** *PDL 9 Microbial mat with enmeshed pyrite octahedrons (100×);* **2.** *Talega 3 Microbial mat with enclosed quartz grains (100×);* **3.** *PDL 6 Microbial mat in cross polarized light (100×);* **4.** *Talega 10 Linear array of organic matter (400×);* **5.** *PDL 1 Phormidium-like filaments attached to quartz grain (400×);* **6.** *Talega 8 Algal-size filament with two empty and one filled cell (100×);* **7.** *Talega 4 Fungal-like spores (400×);* **8.** *Talega 7 Mineralized algal (?) cysts (400×);* **9.** *Talega 8 Pellet-shaped microfossils (40×);* **10.** *Talega 4 Pyrite framboids (400×);* **11.** *PDL 5 Crack filled with hematite (100×);* **12.** *PDL 6 Siderocapsa-like capsular structure with internal iron oxide spheres (100×). See note on p. 113.*

PRELIMINARY ANALYSIS OF PIEDRA DE LUMBRE AND TALEGA CANYON CHERTS

Table 1. *Megascopic data. See note on p. 113.*

Sample No.	Field No.	Field Lithology	Rock Classification	Color of Thin Section	Megascopic Structures and Concentrations of Opaque Minerals
PDL 1	1 Area 1	gray chert	quartz wacke?*	pale brown	opaques concentrated in organic-rich areas
PDL 2	2 Area 5	fractured black chert	chert arenite	pale brown	opaques as elongate arrays
PDL 3	3 Area 6	white chert	chert arenite	gray	large cracks, colorless laminations
PDL 4	4 Area 6	gray chert	chert arenite	brown	organic laminations
PDL 5	5 Area 6	iron-stained sandstone	quartz arenite	orange brown	
PDL 6	6 Area 8	white sandstone	quartz wacke?*	pale brown	
PDL 7	7 Dr. Ezell's (probably Area 6)	gray chert	chert arenite	brown	colorless laminations
PDL 8	8 Area 6	iron-stained sandstone	quartz wacke	pale brown	opaques in patches
PDL 9	9 Area 6	black chert	chert arenite	brown	convoluted laminations
PDL 10	10 Area 6	waxy chert	chert arenite	orange brown	convoluted laminations
PDL 11	Piedra de Lumbre (Dr. Ezell's)	chert	chert arenite	pale brown	convoluted laminations
Talega 1	CA-SDI 13655, surface 5m SW of SW corner of TU2	dark gray chert	chert arenite	pale brown	
Talega 2	CA-SDI 13655, in drainage S of Datum A	dark gray chert	quartz arenite	pale brown	
Talega 3	CA-SDI 13655, surface 1m S of S wall of TU3	dark gray chert	quartz arenite	brown	
Talega 4	CA-SDI 10003, R. Schultz sample, surface, provinance unknown	dark gray chert	chert arenite	brown	convoluted laminations
Talega 5	CA-SDI 13655, STP 9 20-40 cm	dark gray chert	chert arenite	orange brown	
Talega 6	CA-SDI 13655, STP 18 0-20 cm	dark gray chert	quartz wacke	pale brown	
Talega 7	CA-SDI 13655, TU2, 20-30 cm	dark gray chert	chert arenite	orange brown	
Talega 8	CA-SDI 13655, TU2, 10-20 cm	dark gray chert	chert arenite	very pale brown	
Talega 9	CA-SDI 13655, TU1 20-30 cm	dark gray chert	quartz arenite	pale brown	
Talega 10	CA-SDI 13655, TU1 50-60 cm	dark gray chert	chert arenite	pale brown	

* Clay masks microquartz.

Table 2. *Silica content. (Incorporates data from Dr. Girty, in Sikes et al., 2006.) (Abbreviations: A, abundant; CQ, chalcedony; MQ, microquartz; PQ, polycrystalline quartz; R, Rare; S, some; SQ, strained quartz; TQ, Transported quartz). See note on p. 113.*

Sample No.	TQ%-CQ%-MQ%-(clay%)	Quartz Grain Shape, Size, Sorting	Quartz Lithic Fragments	Matrix Composition	Accessory Minerals (< 1% of clasts)	Siliceous Sedimentary Structures
PDL 1	30-0-10-(60)	A-SA-SR <0.01-1mm unsorted	PQ 4%, "ash"	MQ (R), clay	hornblende, sanadine/orthoclase, zircon	resembles stretched geopetal fillings
PDL 2	9-1-90	A-SA-SR <0.01-1.2mm unsorted	PQ 1%	MQ	apatite?, zircon	monocot silica cell
PDL 3	10-5-85	A-SA-SR <0.01-1mm unsorted	PQ <1%, euhedral crystal	MQ	hornblende, rutile	strong & faint laminations, columnar structures
PDL 4	40-0-60	A-SA-SR <0.01-1.5mm unsorted	PQ 10%, "ash", SQ	MQ	hornblende, euhedral zircon	none
PDL 5	90-0-0	A-SA-SR 0.01-1.1mm unsorted	PQ 3%, MQ clast, mylonite	clay	biotite, chlorite?, epidote, hornblende, pyroxene, muscovite, rutile	none
PDL 6	15-10-0-(75)	A-SA-SR <0.01-1.2mm unsorted	PQ <1%, SQ, mylonite, MQ clast	clay	epidote, hornblende, magnetite?, zircon	geopetal fillings, rigid wall structures
PDL 7	10-20-70	A-SA-SR <0.01-0.9mm unsorted	PQ <1%,	MQ	euhedral zircon	laminations of MQ & CQ
PDL 8	90-0-0-(10)	A-SA-SR 0.01-0.9mm unsorted	PQ 10%, MQ clast, mylonite	clay	hornblende, sphene, zircon	none
PDL 9	15-1-80	A-SA-SR <0.01-1.5mm unsorted	PQ <1%, SQ	MQ	none	none
PDL 10	3-10-87	A-SA-SR <0.01-0.7mm unsorted	none	MQ	euhedral zircon, pyroxene	laminar structures, chert flow structures
PDL 11	8-2-90	A-SA-SR <0.01-1mm unsorted	PQ <1%, "ash", SQ	MQ	none	none
Talega 1	15-10-70-(5)	A-SA-SR <0.01-1.8mm unsorted	PQ <1%, "ash", SQ	MQ, clay	epidote, pyroxene, zircon	none

Table 2. *(Continued).*

Sample No.	TQ%-CQ%-MQ%-(clay%)	Quartz Grain Shape, Size, Sorting	Quartz Lithic Fragments	Matrix Composition	Accessory Minerals (< 1% of clasts)	Siliceous Sedimentary Structures
Talega 2	60-0-35-(5)	A-SA-SR-R 0.01-2mm highly unsorted	PQ 10%, "ash", SQ	bimodal MQ, clay	hornblende, magnetite, tourmaline, euhedral zircon	none
Talega 3	50-5-45	A-SA-SR 0.01-1.3mm unsorted	PQ 10%, SQ	MQ	calcite, chlorite, hornblende, pyroxene, zircon	none
Talega 4	19-1-80	A-SA-SR 0.01-1.9mm unsorted	PQ 20%, "ash", SQ	MQ	chlorite, hornblende, pyroxene	none
Talega 5	40-0-60	A-SA-SR <0.01-2mm unsorted	PQ 1%, SQ	MQ	chlorite, pyroxene, rutile	none
Talega 6	80-1-10-(9)	A-SA-SR <0.01-1.4mm unsorted	PQ 1%, mylonite, "ash"	MQ, clay	biotite, chlorite, hornblende, pyroxene, zircon	none
Talega 7	30-10-55	A-SA-SR <0.01-1.0mm unsorted	PQ<1%, "ash", SQ	bimodal MQ	pyroxene, tourmaline?	MQ pisolites
Talega 8	40-5-55	A-SA-SR <0.01-1.7mm unsorted	PQ 4%, "ash", MQ clast, SQ	bimodal MQ	biotite, chlorite, plagioclase, pyroxene, zircon	CQ spheres; CQ laminations
Talega 9	70-1-15	A-SA-SR <0.01-3mm unsorted	PQ 10%, "ash", SQ	bimodal MQ	chlorite, hornblende, pyroxene, rutile, zircon	none
Talega 10	49-0-50-(1)	A-SA-SR <0.01-1.9mm unsorted	PQ 1%, "ash", SQ	bimodal MQ, clay (one side of slide only)	hornblende, olivine?, rutile, zircon	none

Table 3. *Silica and iron types in cracks, veins, and vugs (Abbreviations: AQ, authigenic quartz; CQ, chalcedony quartz; MQ, microquartz; pAQ, authigenic quartz perpendicular to crack; rAQ, authigenic quartz arbitrary orientation? See note on p. 113.*

Sample No.	Structure	Silica Types	Iron Types	Other
PDL 1	veins	none	none	none
PDL 2	cracks, veins	MQ, CQ, pAQ, rAQ	iron oxide	none
PDL 3	cracks, vugs, veins	MQ, CQ, pAQ, rAQ	none	vug cuts crack; full crack cuts empty crack
PDL 4	cracks, veins, vugs	MQ, AQ	none	MQ flow structure
PDL 5	cracks, veins	none	iron oxide	none
PDL 6	none	none	none	none
PDL 7	cracks	MQ, CQ, AQ	none	none
PDL 8	none	none	none	none
PDL 9	cracks	MQ, CQ	iron oxide	none
PDL 10	cracks, vugs, veins	MQ, CQ, AQ	iron oxide	crack with iron oxide cuts vug; 3 generations of silica
PDL 11	cracks	spherulitic CQ	iron oxide	none
Talega 1	cracks, vugs, veins	MQ, CQ, AQ, spherulitic CQ	iron oxide	crack with silica cuts vug
Talega 2	cracks, veins	MQ, AQ	none	none
Talega 3	cracks, veins	MQ, AQ	iron oxide	crack cuts crack
Talega 4	cracks, veins	CQ	iron oxide	organic matter in cracks
Talega 5	cracks, veins	MQ	iron oxide?	none
Talega 6	cracks, vugs, veins	bimodal MQ, CQ	none	none
Talega 7	cracks, veins	MQ, CQ, pAQ, rAQ	iron oxide	none
Talega 8	cracks, vugs, veins	MQ, pAQ, rAQ, spherulitic CQ	none	dissolved organic substance in crack
Talega 9	none	none	none	none
Talega 10	vug	pAQ, rAQ	none	none

Table 4. *Organic matter types and distribution. See note on p. 113.*

Sample No.	Abundance and Distribution	Organic Types and Sedimentary Structures
PDL 1	S?*, dispersed	cf. *Phormidium* attached to quartz grain, filaments, dissolved organic substance
PDL 2	A, dispersed	colonial algae?, spores?, microbial mats, filaments
PDL 3	S-R, dispersed	filaments, microbial mats
PDL 4	A	bifurcating filaments, cells, microbial mats
PDL 5	none	none
PDL 6	none	?
PDL 7	A	microbial mats at acute angle from inorganic laminations
PDL 8	none	none
PDL 9	A	microbial mat
PDL 10	A, dispersed; also in veins	microbial mats
PDL 11	S-A, dispersed	spore, filaments, pellets, algal spheres, disrupted microbial mats
Talega 1	S, dispersed	microbial mat, filaments, coats vugs
Talega 2	S, dispersed	microbial mats
Talega 3	A	microbial mats
Talega 4	S-A, dispersed	fungal? spore, degraded filament and algal spheres,
Talega 5	S-A dispersed	microbial mat, filaments
Talega 6	R, dispersed	linear arrays
Talega 7	S, dispersed & clumped	mat?, degraded filaments, inside cracks, dissolved organic substance
Talega 8	R, dispersed	mat?, linear arrays, pellet
Talega 9	R, clumped	degraded pellet?, filament?
Talega 10	S, dispersed	cyst, degraded filament, pellet?, linear array

* Clay masks microquartz.

Table 5. *Iron content (Abbreviations: A, abundant; S, some; R, rare).*

Sample No.	Hematite	Amorphous Iron Oxide	Iron oxide distribution & abundance	Iron Oxide Biominerals	Amorphous Pyrite	Pyrite distribution & abundance	Pyritized Biominerals
PDL 1	hexagons	linear distribution	R	microbial forms	massive	S, localized, in veins	framboids in strings
PDL 2	hexagons	present	A, in linear cracks, widespread	cf. *Siderocystis*; microbial forms	present	A, interstitial, in vugs & veins	bimodal framboids
PDL 3	hexagons	none	R	none	massive	A, dispersed	bimodal framboids, octahedrons
PDL 4	none	present	R, associated with cracks	microbial forms (spheres, tapered)	massive	R, dispersed in organic matter, localized in veins	octahedrons
PDL 5	hexagons	present	A, interstitial	cf. *Leptothrix discophora* holdfasts; filaments, spheres	none	none	none
PDL 6	hexagons	present	R	cf. *Siderocapsa* capsule	none	R	octahedrons, microbial forms
PDL 7	hexagons	present	R	spheres	massive	cubes	
PDL 8	hexagons	present, oxidized framboids	A, interstitial	cf. *Leptothrix discophora* holdfasts			(oxidized framboids)
PDL 9	hexagons	present	R, localized, along cracks	cf. *Siderocystis*	massive	A	octahedrons in organic matter; microbial forms, framboids
PDL 10	hexagons, non-crystalline	present, oxidized framboids	A, interstitial, along cracks	cf. *Leptothrix discophora* holdfasts	?	none	(oxidized framboids)
PDL 11	none	present	R, in cracks	none	massive	none	none
Talega 1	none	present, oxidizing framboids	R, localized, in vug and crack	none	none	R, dispersed, cube	octahedrons, framboids
Talega 2	none	none	none	none	massive	R, dispersed	octahedrons in organic matter
Talega 3	hexagons	present	S, in cracks & veins	microbial forms	massive	R, dispersed	octahedrons in organic matter
Talega 4	hexagon	present	S, in cracks & veins	cf. *Siderocystis*	massive	S, in veins & laminae	octahedrons in organic matter, framboids

Table 5. *(Continued).*

Sample No.	Hematite	Amorphous Iron Oxide	Iron oxide distribution & abundance	Iron Oxide Biominerals	Amorphous Pyrite	Pyrite distribution & abundance	Pyritized Biominerals
Talega 5	crystalline	present	R, in chert clast, in cracks & veins	none	none	R, dispersed	octahedrons in organic matter, framboids
Talega 6	hexagons	present	R, in cracks	filamentous and spherical microbial forms	massive	R	in organic matter, filamentous microbial forms
Talega 7	none	present	S, interstitial, in vug & cracks	filamentous and spherical microbial forms, capsules	none	R	framboids
Talega 8	none	present	R, along a cell	microbial forms	massive	R	octahedrons in organic matter, framboids
Talega 9	hexagons	none	R	filamentous microbial forms	massive	R, dispersed	in organic matter
Talega 10	hexagons	none	R	filamentous microbial form	massive	R	octahedrons, framboids, filamentous microbial forms

Note: *See Pigniolo (1992, 1994) for PDL and Talega outcrop and site numbers referenced in Figures 6 and 7, as well as Tables 1–5, above.*

REFERENCES

Abbott, P.L., ed., 1985, On the Manner of Deposition of the Eocene Strata in Northern San Diego County: San Diego Association of Geologists Annual Field Trip Guidebook, 98 p.

Abbott, P.L, 1999, The Rise and Fall of San Diego: San Diego, CA, Sunbelt Pubs., p. 75.

Atwater, T., 1970, Implications of Plate Tectonics for the Cenozoic Tectonic Evolution of Western North America: Geological Society of America Bulletin, v. 81, p. 3513-3536.

Behl, R.J., 2011, Chert spheroids of the Monterey Formation, California (USA): early-diagenetic structures of bedded siliceous deposits: Sedimentology, v. 58, p. 325-351.

Berry, R.W., 1999, Eocene and Oligocene Otay-type waxy bentonites of San Diego County and Baja California: Chemistry, mineralogy, petrology and plate tectonic implications: Clays and Clay Minerals, v. 47, p. 70-83.

Blank, C.E., Cady, S.L, and Pace, N.R., 2002, Microbial composition of near-boiling silica-depositing thermal springs throughout Yellowstone National Park: Applied and Env. Microbiology, v. 68, p. 5123-5135.

California Geological Survey, Online digital 7.5' geologic quadrangle maps: http://www.conservation.ca.gov/cgs/rghm/rgm/Pages/preliminary_geologic_maps.aspx

Cranham, G.T., 1985, General geology of the Las Pulgas Canyon area, Camp Pendleton Marine Base, California: B.S. Thesis, San Diego State University, Dept. Geological Sciences, 38 p.

Cranham, G.T., Camilleri, P.A., and Jaffe, G.R., 1994, Geologic overview of the San Onofre Mountains, Marine Corps Base Camp Pendleton, San Diego County, California: republished in Rosenberg, P.S., ed., 2010, Geology and Natural History of Camp Pendleton, 2nd Edition: San Diego Assoc. Geologists, Sunbelt, San Diego, Calif., p. 13-28.

Elliott, W.J., and Berry, R.W., 1991, Bedding plane clay seams: A new look at an old nemesis, in P.L. Abbott and W.J. Elliott (eds.): Environmental Perils San Diego Region, San Diego, San Diego Assoc. Geol. for Geol. Soc. America Ann. Mtg., p. 165-171.

Folk, R.L., 2005, Nannobacteria and the formation of framboidal pyrite: Textural evidence: Jour. Earth Syst. Sci., v. 114, p. 369-374.

Furu, E.J., 1982, Sedimentary Geology of a Portion of Southern Camp Pendleton: M.S. Thesis, San Diego State University, 136 p.

Gallegos, Dennis, and Carrico, Richard, 1984 Windsong Shore Data Recovery Program for Site W-131, Carlsbad, California. Unpublished technical report on file at the South Coastal Information Center, San Diego State University.

Guidry, S.A., and Chafetz, H.S., 2002, Factors governing subaqueous siliceous sinter precipitation in hot springs: examples from Yellowstone National Park, USA: Sedimentology, v. 49, p. 1253–1267.

Hem, J.D., 1989, Study and Interpretation of the Chemical Characteristics of Natural Water: U.S. Geological Survey Water-Supply Paper, 2254, 263 p.

Hinman, N.W., and Walter, M.R., 2005, Textural preservation in siliceous hot spring deposits during early diagenesis; examples from Yellowstone National Park and Nevada, U.S.A.: Jour. Sed. Res., v. 75, p. 200–215.

Jennings, C.W., 1977, Geologic Map of California: California Division of Mines and Geology, Geologic Map No. 2, approximate map scale 1 inch = 12 miles.

Jennings, C.W., 1994, Fault Activity Map of California and Adjacent Areas, with locations and ages of recent volcanic eruptions: California Division of Mines and Geology, Geologic Map No. 6, approximate map scale 1 inch = 12 miles.

Kyle, Jennifer, and Schroeder, Paul, 2005, Evidence of microbial framboidal pyrite formation in a terrestrial hot spring (abs.): Geol. Soc. America, Abs. with Programs, v. 37, p. 205.

Larsen, E.S., Jr., 1948, Batholith and Associated Rocks of Corona, Elsinore, and San Luis Rey Quadrangles Southern California: Geological Society of America Memoir 29, 182 p., map scale 1 inch = 2 miles.

Legg, M.R., Nicholson, Craig, Goldfinger, Chris, Milstein, Randall, and Kamerling, M.J., 2004, Large enigmatic crater structures offshore Southern California: Geophys. Jour. Int., v. 159, p. 803–815.

Maclean, L.C.W., Tyliszczak, T., Gilbert, P.U.P.A., Zhou, D., Pray, T.J., Onstott, T.C., and Southam, G., 2008, A high-resolution chemical and structural stuy of framboidal pyrite formed within a low-temperature bacterial biofilm: Geobiology, v. 6, p. 471–480.

Murray, J., and Fenenga, R., 1981, An Archaeological Survey of an Inland Portion of Joseph H. Pendleton Marine Corps Base, San Diego County, California: Costa Mesa, Archaeological Associates.

Pigniolo, A.R., 1992, Distribution of Piedra de Lumbre "chert" and hunter-gatherer mobility and exchange in Southern California: MS Thesis, San Diego State Univ., Dept. Anthropology, 426 p.

Pigniolo, A.R., 1994, The distribution of Piedra de Lumbre "chert" in the archaeological record of Southern California: Proc. Soc. California Arch., v. 7, p. 191–198.

Robbins, E.I., and Hayes, M.A., 1996, What's the red in the water?, What's the black on the rocks?, What's the oil on the surface?: http://pubs.usgs.gov/publications/text/Norriemicrobes.html

Robbins, E.I., and Norden, A.W., 1994, Microbial oxidation of iron and manganese in wetlands and creeks of Maryland, Virginia, Delaware, and Washington, D.C., in Chiang, S.-H., ed., Pittsburgh Coal Conference Proceedings, Coal--Energy and the Environment, Vol. 2, p. 1154–1159.

Sikes, Nancy, Wesson, Alex, and Hunt, Kevin, 2006, Archaeological Evaluation and Eligibility of Investigation of Site CA-SDI-13,655: Contract Report N68711-04-D-3621, for U.S. Navy Southwest Division, Naval Facilities Engineering Command , 109 p. and appendices.

Soil Survey Staff, 1999, Soil Taxonomy, 2nd ed.:U.S. Dept. Agriculture, NRCS, Agricultural Handbook 436, 869 p.

Stuart, C.J., 1975, The San Onofre Breccia in the Camp Pendleton Area, California: in Ross, A., and Dowlen, R.J., eds., Studies on the Geology of Camp Pendleton, and Western San Diego County, California: San Diego Association of Geologists Annual Field Trip Guidebook, p. 15–27.

Watts, S.H., 2006, A petrographic study of silcrete from inland Australia: Jour. Sedimentary Res., v. 48, p. 987–994.

Weber, F.H., Jr., 1963, Geology and Mineral Resources of San Diego County, California: California Division of Mines and Geology, County Report 3, 309 p., Plate map scale 1"=2 miles.

Weber, F.H., Jr., 1982, Recent Slope Failures, Ancient Landslides, and Related Geology of the North-Central Coastal Area, San Diego County, California: California Division of Mines and Geology, Open-File Report 82-12-LA, 62 p., Plate 1, map scale 1 inch = 2,000 feet.

Wilson, K.L., 1972, Eocene and Related Geology of a Portion of the San Luis Rey and Encinitas Quadrangles, San Diego County, California [M.A. Thesis]: University of California, Riverside, 135 p., map scale 1 inch = 2,000 feet.

Wopfner, H., 1983, Environment of silcrete formation: a comparison of examples from Australia and the Cologne Embayment, West Germany: Geol. Soc. London Spec. Pub., v. 11, p. 151–158.

Yerkes, R.F., McCulloh, T.H., Shoellhamer, J.E., and Vedder, J.G., 1965, Geology of the Los Angeles Basin California – an Introduction: United States Geological Survey Professional Paper 420-A, 57 p.

York, A., and Wahoff, T., 2009 Archaeological Survey and Evaluation of Piedra De Lumbre Quarry (CA-SDI-10,008/10,708, Marine Corps Base Camp Pendleton Phase 1 Investigations. EDAW/AECOM, San Diego. On file at the U.S. Department of the Navy, NAVFAC Southwest, San Diego.

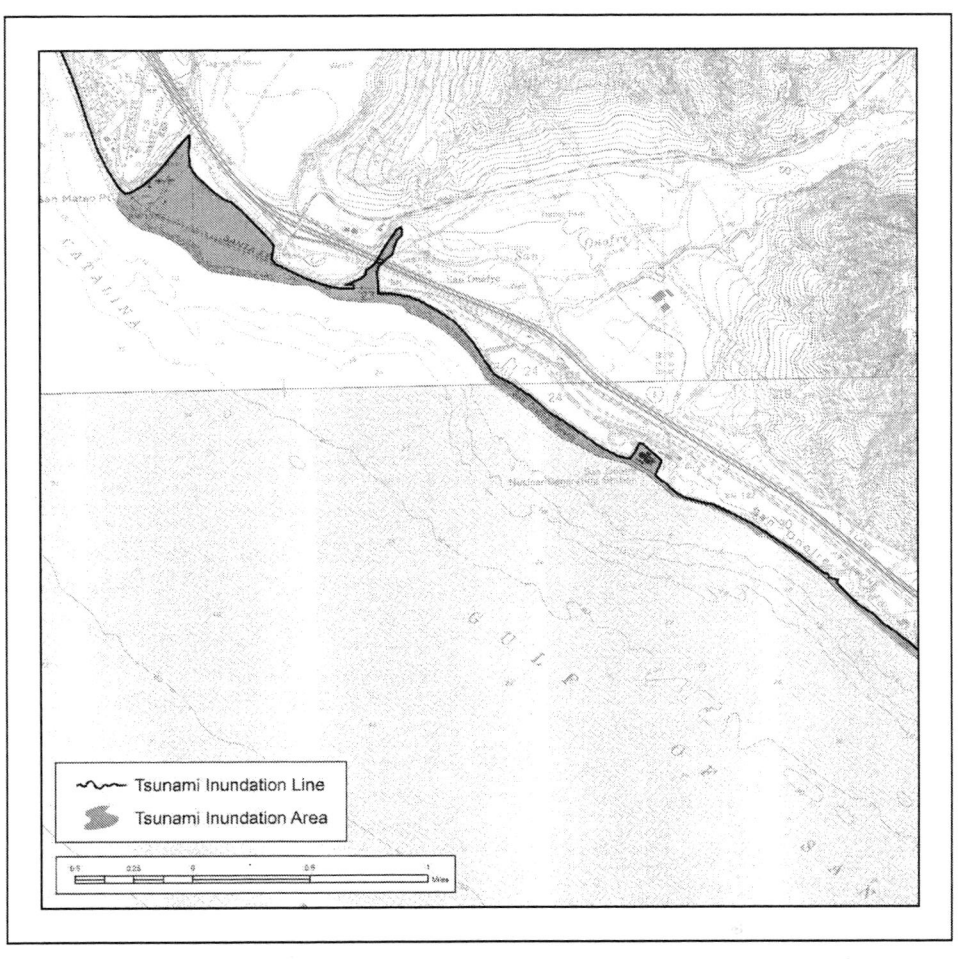

**Portion of San Onofre Bluff Quadrangle,
Tsunami Inundation Map for Emergency Planning, June 1, 2009.**
*(http://www.conservation.ca.gov/cgs/geologic_hazards/Tsunami
/Inundation_Maps/SanDiego/Pages/SanDiego.aspx)*

San Luis Rey River, Oceanside Harbor, March 10, 2008.
Photo: Woodrow L. Higdon, Geo-Tech Imagery.

LAWRENCE CANYON FAULT, OCEANSIDE, CALIFORNIA

William J. Elliott, Consulting Engineering Geologist,
P.O. Box 541, Solana Beach, California 92075

Dr. Monte Marshall
235 Quince Street, San Diego, CA 92103

INTRODUCTION

The Lawrence Canyon fault is located in northwestern Oceanside, California (**Figure 1**). It was visited by San Diego Association of Geologists on April 25, 1981 (Dowlen, 1981, p. 161, Stop 5). For teaching purposes, this is one of the best fault exposures in coastal San Diego County (**Photo 1**).

Pedestrian access is from the cul-de-sac at the north end of McNeil Street, Oceanside, CA. From I-5, exit Mission Avenue west to North Horne Street. Turn north to Civic Center Drive and then go east. Immediately after crossing over I-5 (where Civic Center Drive become Bush Street) turn north on McNeil Street and park at the dead end. From here, walk down-canyon approximately 0.26 mile (about 7 minutes) along the old A.C.-paved quarry access road (be sure to detour east around the very large washout). Google Earth GPS coordinates are: 33° 12.3' North Latitude and 117° 22.7' West Longitude.

Aggregate was mined from San Onofre Breccia by H. W. Rohl Company during the 1950s to provide sand and gravel for local construction projects. Products from this 100 acres site (United States Department of Agriculture, 1953) included, "Concrete and fill sand; crushed gravel (standard sizes rock for use as aggregate; minus 3/16-in. for blacktop)." Weber, 1963, p. 248, Pl. 1, Sequential Catalog No. 403.

LAWRENCE CANYON FAULT

Located on the east side of Lawrence Canyon, latest Miocene/earliest Pliocene (Howard, 1982, p. 1) San Mateo Formation sandstone is in ~N10°W, ~56° E fault contact with lower and middle Miocene (Stuart, 1975, p. 15), San Onofre Breccia. Apparent vertical separation observed at the outcrop is at least 40 feet, down to the east.

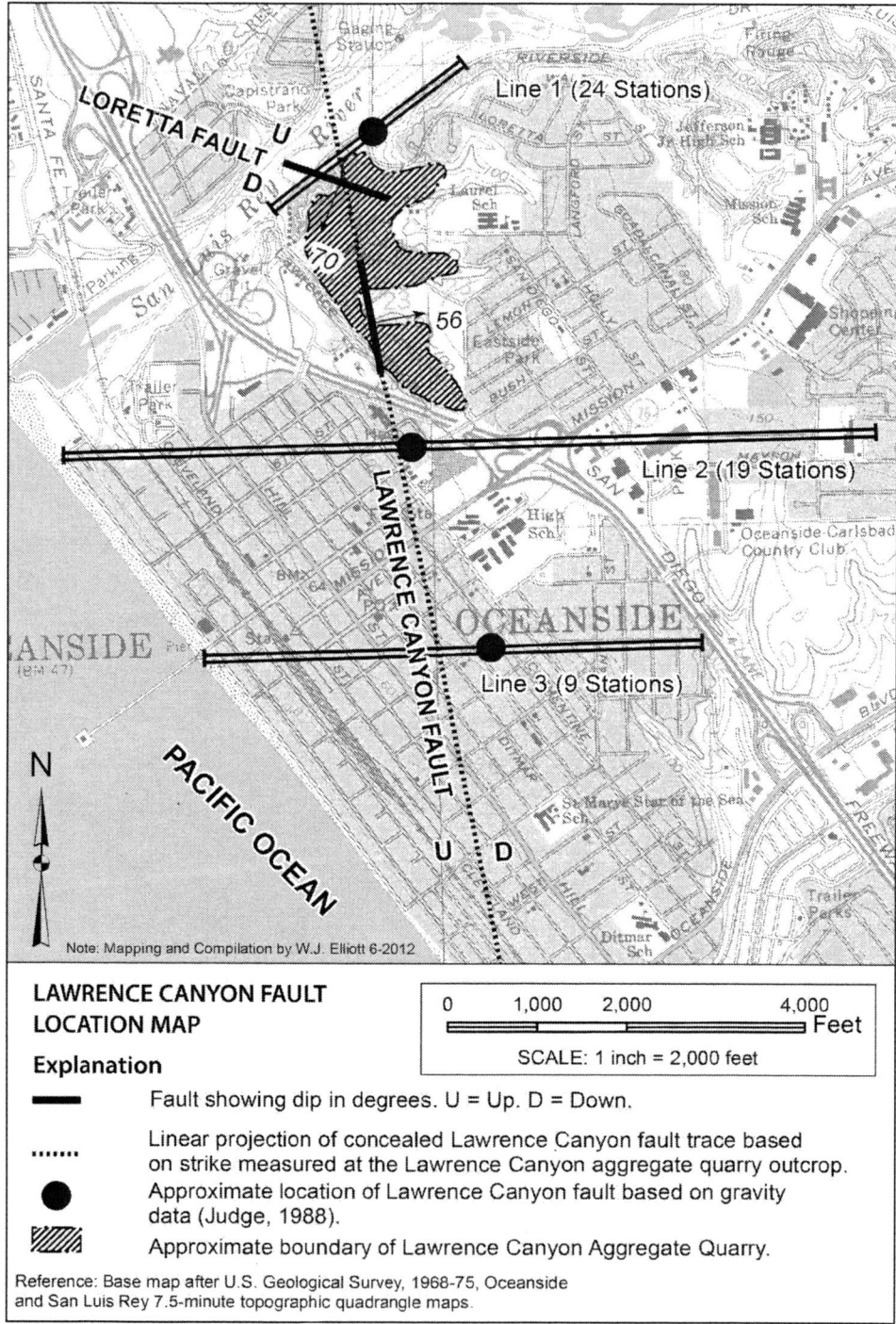

Figure 1. *Location map.*

Now, without giving all the fun stuff away, here are some questions you can answer for yourself when you visit the fault. First, however, take a good look at **Photo 2**. Can you see the narrow, parallel/sub-parallel linear groves and ridges that are roughly aligned with the pencil? Look carefully and you should also be able to see a few thin scratch marks — commonly known as striations. Also, parallel/sub-parallel to the pencil, see if you can pick out several broad, wavy mullion undulations on the slickensided fault surface. The most recent fault movement created these telltale artifacts of San Mateo Formation sliding relatively downward against San Onofre Breccia.

Rake. Now, here is where you come in. What is the rake of these lineations (or *skid-marks*) created by the most recent fault movement? Whoa! This takes me back to my structural geology class — and was that really an F– I earned in Dr. Threet's class? Do you remember that rake is measured in the plane of the fault? Do you also remember that rake is the angle measured between a horizontal line drawn in the plane of the fault (the strike of the fault plane) and the slickenlines?

All you need to do is establish a horizontal line across the fault plane, which you could do with a small hand level (or Brunton compass), and then with a protractor (or Brunton compass) measure the angle between the horizontal line and the slickenlines. And, *voilà*, you now know the rake!

If you were working with a pure strike-slip (lateral) fault, say California's San Andreas, the rake should be 0°. By contrast, a pure dip-slip fault should have a rake of 90°. So, with these end members in mind, what rake angle might you expect to find for a fault with oblique-slip? Now, here's the test. For the Lawrence Canyon fault, what is the rake and apparent sense of slip?

Conventional wisdom has it that, one way to determine which way the missing (eroded) fault block moved is to gently slide your hand parallel to the lineations exposed on the fault surface. The direction that feels smoothest is the direction the missing block moved; sliding your hand in the opposite direction should feel relatively rougher. More recent data, however, shows that this smoothness test can be ambiguous.

These asymmetrical artifacts of fault surface architecture are oftentimes referred to as chatter marks. Look carefully again at **Photo 2**. Along an imaginary line above the pencil, look carefully for chatter marks. Do you see how the San Mateo Formation rode up and over the long smooth ridges? At the bottom of many ridges you will find a break-away step.

If you move your hand gently down this surface it will feel relatively smooth because your hand does not "catch" as it slides easily over the abrupt steps. Conversely, if you move your hand up this surface it will feel rough as it "catches" on each break-away step.

Plunge. Now that your brain has been completely scrambled with long forgotten classroom trivia, lets go one step further and a see if you can remember how to determine the plunge of the slickenlines, as well as their compass direction. Give up?

First visualize a vertical plane passing through one of the narrow ridges or groves in the fault plane. Now visualize a horizontal line in this vertical plane. The angle between the horizontal line and the slickenline is the angle of plunge. The compass direction leading away from the apex of this angle is the bearing of the plunge (Billings, 1960, pp. 128–129).

A Unique Line in Space. Now, having determined the plunge angle and its compass direction, you have defined a unique line in space along which the fault surfaces moved — one past the other. That is, along the line B–C in **Figure 2**.

You are on your own from here. It will be up to you to climb a short distance up into the lower reaches of the Lawrence Canyon fault exposure and make these measurements for yourself. After all, isn't being in the great out-of-doors and fieldwork what attracted most of us to geology in the first place? Go ahead, you can do it!

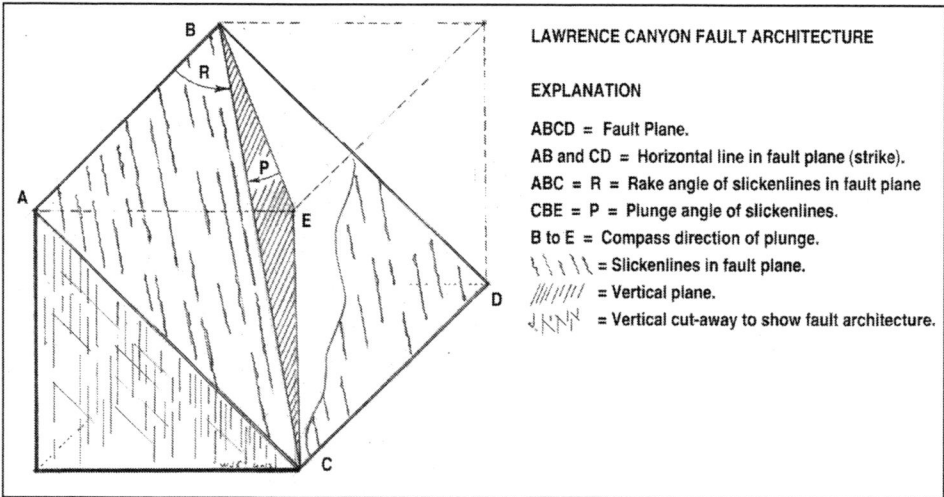

Figure 2. *Fault Architecture.*

INTERPRETATION OF GRAVITY DATA

Figure 1 shows the location of three gravity lines run by Judge (1988) for his senior thesis using a Texas Instruments Worden gravity meter. Line 1 strikes N40°E, whereas lines 2 and 3 strike E-W and are nearly strike-normal to the Lawrence Canyon fault. Line lengths vary from about 3,000 to 9,000 feet. Station spacing ranges from about 100 to 665 feet.

For this study, geologic basement was assumed to have an average density of 2.7 g/cc (grams per cubic centimeter) and surficial sedimentary formations (San Onofre Breccia and San Mateo Formation sandstone were assumed to have a combined average density of 2.3 g/cc. The resulting density contrast is 0.4 g/cc. (Judge, 1988.)

Interpretation of "black-box" gravity data is as much an art as it is a science. Choosing the correct subsurface mathematical architectural model, and correct densities for each element in the model have a direct affect on the final interpretation. However, even before the investigator gets to this part of the study, raw field measurements must be "reduced" by making corrections for, 1) elevation (free-air and Bouguer), 2) local variations in surrounding terrain, 3) latitude, 4) earth-tides, and 5) instrument drift (Dobrin, 1960, pp. 229–235).

The only known exposure of the Lawrence Canyon fault is in the abandoned Lawrence Canyon aggregate quarry. Assuming a constant strike, the dotted line (**Figure 1**) passing through the strike/dip symbol in the quarry represents the projected subsurface fault trace. Along gravity Line 2, the interpreted subsurface fault trace is fairly close to the known surface location, whereas for gravity Lines 1 and 3 the interpreted subsurface trace diverges slightly to the east. This departure could be due to a non-linear fault trace — a characteristic more common for dip-slip faults than for strike-slip faults. It could also result from local subsurface geology being more complex than the simple mathematical model used for interpretation.

Apparent vertical separation (throw) on all three gravity lines is down to the east. It increases from about 100 feet along Line 1, to about 800 feet along Line 3. Average apparent depth to basement varies from about 155 feet at line 1 to about 625 feet at line 3. (Judge, 1988, pp. 18–24, and Figure 10.)

SAN ONOFRE BRECCIA

San Onofre Breccia has been described by, Ellis, and Lee, 1919, p. 57; Woodford, 1925; Stuart, 1975, pp. 15–27; Young, and Berry, 1981, p. 33–52; Weber, 1963; and Weber, 1982, p. 24–25, Pl. 1. Lawrence Canyon exposures (**Photo 3**) consist of: light- to medium-brown and orange-brown, well-indurated, poorly sorted (moderately well-graded), clayey/silty, fine- to coarse-grained sandy gravel (breccia) (GC; **Figure 3**). These angular to sub-rounded, matrix- and clast-supported gravels consist principally of a chaotic mixture of metamorphic rocks (mainly Catalina Schist—Woodford, 1925; Stuart, 1975) and some plutonic rocks. For the most part, clast sizes vary from approximately 1/4- to 8-inches in diameter. Bedding is thick and laterally variable (**Photo 4**), but on close examination it appears massive (**Photo 3**).

SAN MATEO FORMATION

San Mateo Formation has been described by, Woodford, 1925, p. 217; Weber, 1963; Barnes, 1973, p. 37–43; Barnes and others, 1981, p. 53–70; and Weber, 1982, p. 25–26, Pl. 1.

Hand samples of this well sorted (poorly graded) formation (**Photo 5**) have been described, using a 10× triplet hand-lens in bright sunlight, as a coarse silt/very very fine sand (ML/SP; **Figure 3**) of nearly uniform grain size. Somewhere in the neighborhood of about 5 to 10% of a disaggregated sample is easily blown away in a slight breeze—suggesting a limited amount of fine-silt/clay-sized particles.

Composition appears to be largely clear, equidimensional (spherical,) sub-angular to sub-rounded quartz grains, hosting ±2% to 5% biotite (flakes), ±1% to 2% vari-colored (buff, ochre, pale orange) grains, and about ±3% to 7% black minerals (**Photo 6**). No lithic fragments were noted. Less than about 0.01% of the sample was attracted to a powerful magnetite magnet. No reaction was observed with 10% hydrochloric acid (HCl).

Color varies in the range of white to light gray and tan with irregular streaks and nodes of discontinuous yellowish- to orangeish-brown weathering-fronts along bedding and unassigned discontinuities (**Photo 6**).

It is massively bedded, poorly indurated (not cemented), and easily friable between thumb and fingers. In water, it disintegrates readily into a slushy mush.

In outcrop, it is reminiscent of the massive silty portion of the San Diego Formation exposed, for example, in near-vertical cuts along the east side of India Street, just east of Lindbergh Field (Hertlein, and Grant, 1944, pp. 46–63; Kennedy, 1975, p. 29).

LORETTA FAULT

About 0.3 mile northerly from the above-described Lawrence Canyon fault, on the west side of the canyon **(Figure 1)**, a small obscure ~N70°W, ~70°S tending fault, herein named the Loretta fault, is present in the quarry wall (Google GPS coordinates: 33° 12.534′ North latitude and 117° 22.739′ West longitude). Offset San Onofre bedding and subtle drag folding **(Photo 7)** suggest approximately 2.5±0.5 feet of down-to-the-south apparent vertical separation. Beware of poison oak if you attempt a close-up, hands-on view.

ACKNOWLEDGEMENTS

We would like to offer a special thanks to Dave Bloom for his patience with our revisions and for his willingness and expertise in computer graphics to improve our figures from 20th century cut-and-paste to 21st century computer-generated masterpieces. Wendy Elliott, as always, patiently supported our efforts as we worked our way through to the final draft!

FIELD IDENTIFICATION PROCEDURES (Excludes particles larger than 3 inches and basing fractions on estimated weights)						GROUP SYMBOL*
COARSE GRAINED SOILS (More than half of material is larger than No. 200 sieve size¹)	GRAVELS (More than half of coarse fraction is larger than No. 4 sieve size)	CLEAN GRAVELS (Little or no fines)	Wide range in grain size and substantial amounts of all intermediate particle sizes			GW
			Predominantly one size or a range of sizes with some intermediate sizes missing			GP
		GRAVEL WITH FINES (Appreciable amount of fines)	Non-plastic fines (for identification procedures see ML below).			GM
			Plastic fines (for identification procedures see CL below).			GC
	SANDS (More than half of coarse fraction is smaller than No. 4 sieve size)	CLEAN SANDS (Little or no fines)	Wide range in grain size and substantial amounts of all intermediate particle sizes			SW
			Predominantly one size or a range of sizes with some intermediate sizes missing			SP
		SANDS WITH FINES (Appreciable amount of fines)	Non-plastic fines (for identification procedures see ML below).			SM
			Plastic fines (for identification procedures see CL below).			SC
FINE GRAINED SOILS More than half of material is smaller than No. 200 sieve size (The No. 200 sieve size is about the smallest particle visible to the naked eye.)	IDENTIFICATION PROCEDURES ON FRACTION SMALLER THAN No. 40 SIEVE SIZE					
	SILTS AND CLAYS Liquid Limit Less Than 50%		DRY STRENGTH (CRUSHING CHARACTERISTICS)	DILATANCY (REACTION TO SHAKING)	TOUGHNESS (CONSISTENCY NEAR PLASTIC LIMIT)	
			None to slight	Quick to slow	None	ML
			Medium to high	None to very slow	Medium	CL
			Slight to medium	Slow	Slight	OL
	SILTS AND CLAYS Liquid Limit Greater Than 50%		Slight to medium	Slow to none	Slight to medium	MH
			High to very high	None	High	CH
			Medium to high	None to very slow	Slight to medium	OH
HIGHLY ORGANIC SOILS			Readily identified by color, odor, spongy feel and frequently by fibrous texture.			Pt

Figure 3. *Unifed Soil Classification Chart. After: United States Department of the Interior, Bureau of Reclamation: 1974, 2nd edition, Figure 7. (Note: All sieve sizes on this chart are U.S. Standard.)*

GROUP SYMBOL*	TYPICAL NAMES	INFORMATION REQUIRED FOR DESCRIBING SOILS
GW	Well graded gravels, gravel-sand mixtures, little or no fines.	Give typical name; indicate approximate percentages of sand and gravel; maximum size; angularity, surface condition, and hardness of the coarse grains; local or geologic name and other pertinent descriptive information; and symbol in parentheses. For undistributed soils add information on stratification, degree of compactness, cementation, moisture conditions, and drainage characteristics. Example: **Silty sand,** gravelly; about 20% hard, angular gravel particles 1/2-in. maximum size; rounded and subangular sand grains coarse to fine; about 15% nonplastic fines with low dry strength; well compacted and moist in place; alluvial sand; (SM).
GP	Poorly graded gravels or gravel-sand mixtures, little or no fines.	
GM	Silty gravels, poorly graded gravel-sand-silt mixtures.	
GC	Clayey gravels, poorly graded gravel-sand-clay mixtures.	
SW	Well graded sands, gravelly sands; little or no fines.	
SP	Poorly graded sands, gravelly sands, little or no fines.	
SM	Silty sands, poorly graded sand-silt mixtures.	
SC	Clayey sands, poorly graded sand-clay mixtures.	
ML	Inorganic silts and very fine sands, rock flour, silty or clayey fine sands with slight plasticity.	Give typical name: indicate degree and character of plasticity, amount and maximum size of coarse grains, color in wet conditions, odor if any, local or geologic name, and other pertinent descriptive information, and symbol in parentheses. For undisturbed soils add information on structure, stratification, consistency in undisturbed and remolded states, moisture and drainage conditions. EXAMPLE: **Clayey silt,** brown; slightly plastic; small percentage of fine sand; numerous vertical root holes; firm and dry in place; loess; (ML).
CL	Inorganic clays of low to medium plasticity, gravelly clays, sandy clays, silty clays, lean clays.	
OL	Organic silts and organic silt-clays of low plasticity.	
MH	Inorganic silts, micaceous or diatomaceous fine sandy or silty soils, elastic silts.	
CH	Inorganic clays of high plasticity, fat clays.	
OH	Organic clays of medium to high plasticity.	
Pt	Peat and other highly organic soils.	

Figure 3. *Unifed Soil Classification Chart (continued).*
* Boundary classifications: Soils possessing characteristics of two groups are designated by combinations of group symbols. For example GW-GC, well graded gravel-sand mixture with clay binder.

GROUP SYMBOL*	LABORATORY CLASSIFICATION CRITERIA			
GW	Use grain size curve in identifying the fractions as given under field identification.	Determine percentages of gravel and sand from grain-size curve. Depending on percentage of fines (fraction smaller than No. 200 sieve size) coarse-grained soils are classified as follows: Less than 5%: GW, GP, SW, SP. More than 12%: GM, Gc, SM, Sc. 5% to 12%: Borderline cases require use of dual symbols	$C_u = \dfrac{D_{60}}{D_{10}}$ Greater than 4 $\quad C_e = \dfrac{(D_{30})^2}{D_{10} \times D_{60}}$ Between 1 and 3	
GP			Atterberg limits above "A" line with PI greater than 7	
GM			Atterberg limits below "A" line, or PI less than 4	Above "A" line with PI between 4 and 7 are borderline cases requiring use of dual symbols.
GC			Atterberg limits above "A" line with PI greater than 7	
SW			$C_u = \dfrac{D_{60}}{D_{10}}$ Greater than 4 $\quad C_e = \dfrac{(D_{30})^2}{D_{10} \times D_{60}}$ Between 1 and 3	
SP			Atterberg limits above "A" line with PI greater than 7	
SM			Atterberg limits below "A" line, or PI less than 4	Above "A" line with PI between 4 and 7 are borderline cases requiring use of dual symbols.
SC			Atterberg limits above "A" line with PI greater than 7	
ML	**PLASTICITY CHART FOR LABORATORY CLASSIFICATION OF FINE GRAINED SOILS**			
CL				
OL	COMPARING SOILS AT EQUAL LIQUID LIMIT — Toughness and dry strength increase with increasing plasticity index.			
MH				
CH				
OH				
Pt				

Figure 3. *Unifed Soil Classification Chart (continued).*

Photo 1. *Northeasterly view of Lawrence Canyon Fault. Brown and orange-brown San Onofre Breccia on left is faulted against white San Mateo Formation on right. Fault dips down to right in photograph. Four large bushes have taken up residence along the fault trace. Photo by W. J. Elliott, 4-12-12.*

Photo 2. *Close-up view of slickensided Lawrence Canyon fault surface. Pencil is approximately parallel/sub-parallel to ridges/groves, striations and mullions (slickenlines or skid-marks). Hand-lens hangs vertical. Note chatter marks above pencil. Photo by W. J. Elliott, 6-1-12.*

Photo 3. *Close-up view of San Onofre Breccia. Incremental scale on Jacob's staff is 1 foot. Photo by W. J. Elliott, 4-12-12.*

Photo 4. *Southeasterly view of apparently nearly-horizontal San Onofre Breccia bedding. Massive white San Mateo Formation sand is exposed on right. Lawrence Canyon fault plane is obscured. Photo by W. J. Elliott, 4-12-12.*

Photo 5. *Overview of rilled and eroded quarry cut-slope in San Mateo Formation sandstone. Note shovel in lower center of outcrop for scale. Photo by W. J. Elliott, 6-1-12.*

Photo 6. *Close-up view of massive San Mateo Formation sandstone. Photo by W. J. Elliott, 6-1-12.*

Photo 7. *Westerly view of Loretta fault in San Onofre Breccia near northern end of Lawrence Canyon aggregate mine. Note drag folding and approximately 2 to 3 feet of down-to-the-left apparent vertical separation. Photo by W. J. Elliott, 4-12-12.*

REFERENCES

Barnes, L. G., 1973, Pliocene Cetaceans of the San Diego Formation, San Diego, California, in Ross, A., and Dowlen, R. J., eds., Studies on the Geology and Geologic Hazards of the Greater San Diego Area, California: San Diego Association of Geologists, San Diego, California, annual field trip guidebook, p. 37–42.

Barnes, L. G., Howard, H., Hutchison, J. H., and Welton, B. J., 1981, The Vertebrate Fossils of the Marine Cenozoic San Mateo Formation at Oceanside, California: in Abbott, P. L., and O'Dunn, S., eds., Geological Investigations of the Coastal Plain, San Diego County, California: San Diego Association of Geologists annual field trip guidebook, April 25, 1981, p. 53–70.

Billings, M.P., 1960, Structural Geology: Prentice-Hall, Inc., Englewood Cliffs, New Jersey, second edition, 514p.

Dobrin, M. B., 1960, Introduction to Geophysical Prospecting: McGraw-Hill Book Company, Inc., New York, 446p.

Dowlen, R. J., 1981, Field Trip Guide to Geologic Features in North-Coastal San Diego County, California: in Abbott, P. L., and O'Dunn, S., eds., Geological Investigations of the Coastal Plain, San Diego County, California: San Diego Association of Geologists annual field trip guidebook, April 25, 1981, p. 157–166.

Ellis, A. J., and Lee, C. H., 1919, Geology and Ground Waters of the Western Part of San Diego County, California: United States Geological Survey Water-Supply Paper 446, 321p., map scale $1'' = 4$ miles.

Hertlein, L. G., and Grant, U. S., IV, 1944, The Geology and Paleontology of the Marine Pliocene of San Diego, California — Part I, Geology: Memoirs of the San Diego Society of Natural History, Volume II, 72p., 18 Plates.

Howard, H., 1982, Fossil Birds from Tertiary Marine Beds at Oceanside, San Diego County, California, with Descriptions of Two New Species of the genera Uria and Cepphus (Aves: Alcidae): Natural History Museum of Los Angeles County, Contributions in Science, Number 341, pages 1–15. (found on line).

Judge, P., 1988, Gravity Survey of a Little Known Fault in Oceanside, California [B.S. Thesis]: Department of Geology, San Diego State University, 33p. (Faculty advisor, Dr. Monte Marshall).

Kennedy, M. P., 1975, Geology of the San Diego Metropolitan Area, California, Section A, Western San Diego Metropolitan Area: California Division of Mines and Geology Bulletin 200, pp. 1–39, map scale $1'' = 2,000'$.

Stuart, C. J., 1975, The San Onofre Breccia in the Camp Pendleton Area, California, in Ross, A., and Dowlen, R. J., eds., Studies on the Geology of Camp Pendleton, and Western San Diego County, California: San Diego Association of Geologists field trip guidebook, p. 15–27.

Unites States Department of Agriculture, 1953, Stereographic black and white aerial photographs, Nos. AXN-9M-191, 192, flown April 14, 1953.

United States Department of the Interior, 1974, Earth Manual, A Water Resources Technical Publication: United States Department of the Interior, Bureau of Reclamation, second edition, 810p.

United States Geological Survey, 1968-75a, Oceanside, California, 7½' Topographic Map, scale 1" = 2,000', contour interval 20'.

United States Geological Survey, 1968-75b, San Luis Rey, California, 7½' Topographic Map, scale 1" = 2,000', contour interval 20'.

Weber, F. H., Jr., compiler, 1963, Geology and Mineral Resources of San Diego County, California: California Division of Mines and Geology County Report 3, 309p., Plate 1 map scale 1" = 2 miles.

Weber, F. H., Jr., 1982, Recent Slope Failures, Ancient Landslides, and Related Geology of the North-Central Coastal Area, San Diego County, California: California Division of Mines and Geology Open-File Report 82-12-LA, 77p., Pl. 1, map scale 1" = 2,000'.

Woodford, A. O., 1925, The San Onofre Breccia, Its Nature and Origin: University of California Publications, Bulletin of the Department of Geological Sciences, v. 15, no. 7, pp. 159–280.

Young, J. M., and Berry, R. W., 1981, Tertiary Lithostratigraphic Variations, Santa Margarita River to Agua Hedionda Lagoon: in Abbott, P. L., and O'Dunn, S., eds., Geological Investigations of the Coastal Plain, San Diego County, California: San Diego Association of Geologists annual field trip guidebook, April 25, 1981, p. 33–51.

**Inlet channel (left) and discharge channel (right),
Agua Hedionda Lagoon, Carlsbad. November 13, 2007.**
Photo: Woodrow L. Higdon, Geo-Tech Imagery.

Beacon's Beach landslide, Leucadia, December 16, 1996.
Photo: Woodrow L. Higdon, Geo-Tech Imagery.

LANDSLIDE GAPS AT BEACON'S BEACH, ENCINITAS, CALIFORNIA
We've all Heard about Seismic Gaps

William J. Elliott

Consulting Engineering Geologist, P.O. Box 541, Solana Beach, California 92075

INTRODUCTION

Landslides abound at Beacon's Beach!

Most of us have heard about "seismic gaps." Well, here is a new one, "landslide gaps."

Located in the community of Leucadia (United States Geological Survey, 1968-75), this popular surfing spot is developed with numerous cliff-edge homes, many of which are protected by a variety of sea cliff armoring schemes. As you wander north and south along this beach, keep a mental tally of the many creative sea cliff armoring schemes.

FAULTS, STRATIGRAPHY AND LANDSLIDES

From the north end of Beacon's Beach parking lot, to a point about 0.35 mile south, dark gray and black Eocene siltstone and claystone sediments are exposed between overlying Quaternary terrace deposits and beach sand below. Two major northeasterly trending faults separate these fine-grained Eocene sediments from Torrey Sandstone to the south and from a variety of sandy Santiago Formation sediments to the north (Elliott, 2001, Figure 1, p. 14)

Three prominent landslides occur within this short distance. First, an historical landslide spans the width of the Beacon's Beach parking lot; its last significant failure was in February, 1987 (Charlie Lough, personal communication). Second, a prominent pre-human history (but geologically young) landslide is situated at the southern end of this short stretch of beach. Third, the June 1996 landslide is sandwiched in the middle, where several cliff-top residences lost rear yard decks and out-buildings. (Elliott, 1996, cover images.)

LANDSLIDE GAPS

Gap No. 1. Landslide gaps lie between each of these three slides. One that can be relatively easily observed lies between the Beacon's Beach and June 1996 slides. Approximately 145 feet south from the southern end of Beacons Beach parking lot recent storms have removed a small patch of vegetation from the sea-cliff **(Photo 1)**. Here, dark gray to black siltstone and claystone sediments are exposed at beach level in the lower sea cliff (Google Earth GPS coordinates: 33° 03.86' North Latitude and 117°18.3' West Longitude).

In **Photo 2**, un-fractured, in-place sediments occur below the head of the rock hammer and white clipboard. In sharp contrast, fractured sediments above the head of the rock hammer show clear evidence of previous movement. The dividing line, a thin clay seam, is a likely candidate slip-surface for the next landslide.

Gap No. 2. A plane of weakness for the landslide gap between the June 1996 and southernmost slide is obscured by surficial slope wash and/or bluff retention structures.

ACTIVE LANDSLIDE

With a little effort, the clay-seam slip surface at the northern end of the southernmost landslide can be found at about head-height above beach level; this slide is still moving. See if you can find where the toe of the slide extends slightly out over underlying sediments. Swipe your finger along the slide surface; did you feel the soft wet gray clay?

LANDSLIDES AND THE FUTURE

While it is all but impossible to predict exactly when the next landslide event will occur, it is possible to delineate where future landslide events might occur. It is with geologic knowledge, such as is described above, that landslide gaps can be identified.

Our hope is that no one will be hurt and property damage/loss will be minimal when these events eventually occur. As Engineering Geologists, we can't tell people what to do or how to live their lives, but we can provide information from which informed decisions can be made where life and limb may be at risk.

ILLUSTRATIONS

Photo 1. *Easterly overview of landslide gap between the Beacon's Beach slide (on left) and the June 1996 slide (on right). Photo by W. J. Elliott, 4-12-12.*

Photo 2. *Close-up view of fractured siltstone and claystone overlying un-fractured siltstone and claystone. Clay seam at head of rock hammer marks location of a potential landslide slip surface. Incremental scale on Jacob's staff is 1 foot. Photo by W. J. Elliott, 4-12-12.*

REFERENCES

Regional and site-specific references can be found in Elliott, 2001.

Elliott, W. J., 1996, Cover Photographs of June 2, 1996, Neptune Avenue (800 block) Landslide, Encinitas, California: San Diego Association of Geologists annual field trip guidebook, three cover photographs with inside cover caption.

Elliott, W. J., 2001, Coastal Landsliding — Leucadia, California, in Stroh, R. C., ed., Coastal Processes and Engineering Geology of San Diego, California: San Diego Association of Geologists annual field trip guidebook, pp. 13–29.

United States Geological Survey, 1968–75, Encinitas, California, 7½' topographic map, scale 1" = 2,000', contour interval 20'.

SHORELINE HISTORY AND MYSTERY SOLANA BEACH, CALIFORNIA

William J. Elliott
*Consulting Engineering Geologist,
P.O. Box 541, Solana Beach, California 92075*

INTRODUCTION

For most of the approximately 1½ miles of Solana Beach shoreline (United States Geological Survey, 1967-75, 1968-75), Bay Point Formation unconformably overlies Torrey Sandstone. Sandwiched within this almost monotonous, yet starkly beautiful shoreline there are as many as three channels carved into middle Eocene Torrey Sandstone and then covered over by late Pleistocene Bay Point Formation terrace deposits. (Hanna, 1926; Strand, 1962; Weber, 1963; Rogers, 1966; and Kennedy, 1975, pp. 23 and 29, Pl. 1A.)

Carter, 1957, observed two channels along the Solana Beach shoreline (p. 83) and provided a cartoon sketch of what is probably the southernmost one on page 85 (see discussion of the Del Mar Shores channel below).

At each of these ancient channels, a bight or cove has been eroded into the otherwise relatively "straight" shoreline. These relatively protected pocket beaches are favorite hangouts for beachgoers. Read on to learn more about the geological aspects of these channels and their popularity as recreational hangouts.

GEOLOGY OF THREE POPULAR BEACHGOERS HANGOUTS

Pill Box. The number one hangout is Pill Box, headquarters for Solana Beach Lifeguards. Here abundant parking, a beach-side park with tot-lot, restrooms, and an easily negotiated ramp down to the sand make this a popular recreation site for folks of all ages and stages in life. Sunbathing, surfing and family gatherings are all popular Pill Box pastimes.

The first, and most easily identified and observed paleo-channel is located at Fletcher Cove, known to local residents as, *Pill Box* (**Photos 1 and 2**) . This enigmatic name is a leftover from World War II, when anti-aircraft gun emplacements and associated support bunkers were present along much of San Diego County's coast-line.

Before we go any further, lets digress for a moment to set the scene. An interesting combination of events dovetailed during the past 100 or so years to expose this channel as we see it today.

First, early/mid 20th century modifications to river and stream channels, as well as construction of Oceanside Harbor jetties starting in 1942, have blocked/slowed the natural net southward transport of *Oceanside Littoral Cell* beach sand (Flick, 1993; Nordstrom, and Inman, 1973).

Second, with beaches south of Oceanside Harbor now starved for sand, a series of major late 20th century storms easily battered and eroded seacliffs at their base. For example, the wet and stormy winters of: 1951–52, 1957–58, 1965–66, 1977–78, 1978–79, 1979–80, 1982–83, 1985–86, 1992–93, 1994–95, 1997–98, and most recently, 2004–05. (http://www.climatestations.com/san-diego/)

Third, years of rules and regulations promulgated by the 1972 voter-authorized California Coastal Commission have led to a *let-nature-have-her-way* permitting philosophy versus allowing timely construction and/or maintenance of sea cliff armor to protect both public and private cliff-top property improvements.

When the author moved to Solana Beach in 1970, the *Pill Box* channel was protected and completely obscured by a substantial shoreward sloping shotcrete face (**Photo 3**). By 1979, this armor was being seriously battered, broken, and eroded, never to be maintained and/or replaced (**Photo 4**).

Located at the west end of Lomas Santa Fe Drive, channel width is about 360 feet. Pocket beach depth is nominally 70 feet (San Diego, County of, 1975). Google Earth GPS coordinates are: 32° 59.5′ North Latitude and 117° 16.46′ West Longitude.

For the most part, these channel deposits are composed of dark gray to black silty/clayey sediments in the center (**Photo 5**). Weathered tan and light brown silty and sandy sediments drape inward toward the channel center on each side (**Photo 6**). Clam and leaf-stem impressions are common in fine-grained sediments toward central and upper portions of the channel (**Photos 7 and 8**).

An eastward extension of this channel was found in May 1993 by Geocon, Inc. while drilling exploratory test borings for the Lomas Santa Fe/Santa Fe Railroad grade separation project (David B. Evans, personal communication, April 16, 2012).

During grade separation excavation, Dr. Tom Demére, San Diego Natural History Museum, collected paleontological samples from the exposed channel. In his opinion, these sediments are of late Pleistocene age, falling in the time frame of approximately 120,000 to 200,000 years BP (personal communication, March 9, 2004).

As luck would have it, this railroad grade separation exposure provided evidence for not only age, but of an approximately 0.2-mile long, roughly east-west trending channel. To date, additional channel exposures are unknown.

Photo 9 shows a well-defined landslide head-scarp, mid-slope bench, and toe on the north side of Fletcher Cove. Two sunbathers lie at the toe of this recent event. Light colored tan and orange Bay Point Formation silty sands, along with dark colored silty/clayey channel deposits have slumped onto the beach. Sadly, resultant sea-cliff retreat is toward the recently refurbished Solana Beach Community Center that sits close to the sea cliff edge. Nature *is* having her way!

Del Mar Shores. The number 2 surfing and sunbathing hangout seems to be at the foot of the public-access stairs at the west end of Del Mar Shores Terrace. Beyond a year-round public parking lot, not too far from the head of the stairs, and summertime lifeguard service, there are no public services.

The Del Mar Shores channel is located approximately 0.65 miles south of Fletcher Cove, immediately south of the Del Mar Shores public beach stairs. Access is from the west end of Del Mar Shores Terrace. Google Earth GPS coordinates are: 32° 58.93' North Latitude and 117° 16.38' West Longitude.

Here, light colored silty sands fill the approximately 520 feet wide channel (San Diego, County of, 1975, 1985a; **Photo 10**). This narrow, rectangularly shaped pocket beach is about 60 feet deep. Riprap has been used to stabilize the lower bluff slope at the northern end, and a concrete seawall is present at the southern end. A crib-wall and shotcrete facings have been installed to stabilize the upper bluff (Bay Point Formation) above the seawall. Scattered vegetation presides on the otherwise unprotected slope above the riprap. THEN and NOW images can be seen in **Photos 11 and 12**.

During an unsuccessful attempt to stabilize the southern portion of the upper sea cliff during spring of 1979, a concrete *Lincoln-logs* crib wall failed catastrophically, scattering concrete *logs* on the beach below (**Photo 13**).

Tide Park. This lesser known cove is regularly visited by surfers and sunbathers.

Several yards north of Tide Beach, tide pools pique the imagination of curious beach visitors. Beyond summertime lifeguard service, and the access stairs, there are no public services.

Known by locals as *Tide Beach*, or *The Cove*, it is located approximately 0.45 mile north of Fletcher Cove at the west end of Solana Vista Drive. Google Earth GPS coordinates are: 32° 59.87′ North Latitude and 117° 16.58′ West Longitude.

Channel width is about 140 feet; pocket beach depth is about 110 feet (San Diego, County of, 1985b). **Photos 14 and 15.**

Since sometime in the early 1970s, the lower reaches of the sea cliff at the back of this pocket beach have been armored against wave attack by bags of concrete stacked one on top of the other, up to the Torrey Sandstone / Bay Point Formation contact **(Photo 16).**

Both 1939 and 1953 black and white stereographic aerial photographs (United States Department of Agriculture, 1939, 1953) show no obvious protection at the back of Tide Park. In fact, 1939 images show a rounded shadow at the back of the cove suggesting the presence of a natural or hand-excavated sea cave. In 1953 two shadows in the lower sea cliff suggest the presence of a second opening.

The origin of these "mystery" shadows may never be known, as sometime between 1953 and the early 1970s, the lower sea cliff was covered and stabilized with *bag-crete*. **Photos 14, 15 and 16.**

From all outward appearances, a Pleistocene channel, similar to those at Fletcher Cove and Del Mar Shores, appears to be a good first approximation for the origin of Tide Park pocket beach. If a Pleistocene channel is incorrect, then the next best possibility would seem to be something akin to extensively fractured and faulted Torrey Sandstone which has been eroded so far back as to create the observed pocket beach.

During the early 1970s, we relative newcomers would listen reverently to old timers' lengthy yarns of sea caves extending all the way to Old Highway 101 (Old Pacific Coast Highway). To me, this seemed a bit of a stretch. While exploring Solana Beach's sea caves with my twin boys **(Photos 3 and 10)**, I'd never been able to walk or crawl further back into the numerous caves further than about a hundred or so feet before my then-skinny body kept me from going further.

Here is where this tall tale is going. It is possible that shadows seen in

1939 and 1953 photographs were actually hand-dug caves/tunnels that may have gone back some long-ago-forgotten and exaggerated distance. Looking back, it does not seem unreasonable to imagine that children and/or adults with nothing better to do had dug into the lower (sandy?) sea cliff until it was decided by others that this was not the safest pastime. Perhaps we will never know the origin of the shadows or the reason for placing *bag-crete* at the back of the cove — but it makes a great story anyway!

SEA CAVES AND SEA CLIFF ARMOR

Once plentiful sea caves, eroded to depths from a few feet to tens of feet, along near-vertical, northeasterly trending faults and fractures (**Photo 17**) are now mostly filled and/or covered with concrete (**Photo 18**). These damp and dark, shadowy caverns were great places to play hide and seek, as well as make-believe reenactments of explorers and pirates.

Now-a-days, besides concrete blocking entrances to sea caves, concrete armor is being used to protect sea cliffs by blunting wave attack at the toe (**Photos 18 and 19**). In years past, wide sandy beaches absorbed storm wave attack during all but the most severe events — that usually coincided with high tides. Today, as thin and scarce remaining supplies of sand drift steadily southward toward La Jolla (Flick, 1993), heroic measures are required to salvage expensive, property-tax-paying bluff-top homes.

It seems like some combination of short groins (placed perpendicular to the shoreline) and/or submerged offshore reefs (placed parallel to the shoreline) would, over the long-term, benefit beachgoers and homeowners alike. These measures would help retain what little sand remains, provide permanent sandy beaches for sunbathers, and reduce the need to armor sea cliffs with concrete against a never ending battle with the restless sea. Give her an inch, and mother nature will take her mile!

ACKNOWLEDGEMENTS

My thanks go to David B. Evans, Geocon, Inc., for his assistance with the Pill Box channel extension location in the Lomas Santa Fe Drive/Santa Fe Railroad grade separation excavation, and to Dr. Tom Deméré for pointing out early San Diego County geo-work done by George Carter. And not to be forgotten, my dear wife, Wendy, who asks quietly and patiently, "Is this going to be the last paper — at least for a while?"

ILLUSTRATIONS

Photo 1. *Northeasterly view of Fletcher Cove. Note dark colored silt/clay channel deposits on right and at toe of landslide. On left, Torrey Sandstone is overlain by Bay Point Formation terrace deposits. Photo by W. J. Elliott, 4-15-2012.*

Photo 2. *Southeasterly view of Fletcher Cove. Note last remains of old shotcrete armor to right of stairs and black channel deposits to right of that. On far right, Bay Point Formation overlies Torrey Sandstone. Solana Beach lifeguard building is perched at the edge of eroding sea cliffs in center of photograph. Photo by W. J. Elliott, 4-15-2012.*

Photo 3. *Beyond my two silly children, (left to right) Trent and Travis, note sloping protective shotcrete face between the sandy beach and chain-link fence, above (Compare with Photo 1). Photo by W. J. Elliott, July 1975.*

Photo 4. *Broken and eroding shotcrete facing on south side of Fletcher Cove. As eventual sea cliff retreat encroached under concrete viewing porch at top of slope it was finally removed for safety reasons (compare with Photo 2). Photo by W. J. Elliott, 1-30-1979*

Photo 5. *Dark gray and black silty/clayey channel deposits on south side of Fletcher Cove. Shotcrete facing on right of stairs is last remnant of old original lower sea cliff armor (compare with Photo 4). Photo by W. J. Elliott, 4-15-2012.*

Photo 6. *Light colored silty/sandy lenses drape into channel on north side of Fletcher Cove. Torrey Sandstone on left. Bay Point Formation overlies channel deposits in center and right of photograph. Modern gray beach sand at bottom of photograph was mechanically placed to provide minimal toe "protection" during stormy winter months. Photo by W. J. Elliott, 4-15-2012.*

Photo 7. *Clam impressions. Photo by W. J. Elliott, 4-19-2012.*

Photo 8. *Leaf impressions. Yellowish stem and branching veins open downward in photograph. Both halves are shown in photograph. Photo by W. J. Elliott, 4-19-2012.*

Photo 9. *Recent landslide above and beyond sun bathers. Dark colored silt/clay channel deposits on right, between landslide and concrete barrier. Bay Point Formation terrace deposits overlie Torrey Sandstone on left of photograph, and channel deposits on right. Compare left side of photograph with Photo 6. Photo by W. J. Elliott, 4-15-2012.*

Photo 10. *Del Mar Shores channel sands. Note gravelly cross-bedded sands on either side of my son, Trent. Top of channel is at top-left of photograph—and appears to slope gently down to the right, away from viewer. Photo by W. J. Elliott, 2-19-1978.*

Photo 11. THEN: *Southerly view of Del Mar Shores pocket beach. Riprap protects northern end and a concrete seawall protects southern end from wave attack. A combination of concrete and brush protect upper sea cliffs. Compare channel sand deposits in lower one third of sea cliff with Photo 10. Photo: W. J. Elliott, 12-22-1978.*

Photo 12. NOW: *Southerly view of Del Mar Shores pocket beach. Compare with Photo 11. Photo by W. J. Elliott, April 26, 2012.*

Photo 13. *Failed attempt to construct a concrete Lincoln-logs crib-wall to protect upper sea cliff Bay Point Formation between sea wall below and condominiums above. Compare with eventual solution, a stepped shotcrete face above near end of sea wall as seen in Photo 12. Photo by W. J. Elliott, 4-28-1979.*

Photo 14. *Northwesterly view of Tide Park pocket beach from beach access stairs. Note bag-crete armor that protects lower sea cliffs on lower right side of image. Bay Point Formation terrace deposits unconformably overlie Torrey Sandstone in middle distance, and appear to "overlie" bag-crete in foreground. Scattered salt-tolerant vegetation provides limited protection to fragile upper cliff face. Photo by W. J. Elliott, 4-15-2012.*

Photo 15. *Southeasterly view of Tide Park beach. Note bag-crete armor at back of pocket beach. Torrey Sandstone cliffs on either side are overlain by Bay Point Formation terrace sands. Photo by W. J. Elliott, 4-15-2012.*

Photo 16. *Close-up view of bag-crete armor on left, Torrey Sandstone on right and Bay Point Formation terrace sands above. Salt-tolerant vegetation provides limited protection to vulnerable terrace sands above armored lower sea cliff. Photo by W. J. Elliott, 4-15-2012.*

Photo 17. *A pair of sea caves north of Pill Box. Note the fairly intact shotcrete facing on the upper bluff above the sea caves. Photo by W. J. Elliott, 10-17-1981.*

Photo 18. *A concrete wall covers sea caves shown in Photo 17. Note the badly deteriorating remains of the shotcrete facing above the concrete wall that covers the two sea caves. This photo was taken at a +2.3 tide. Note the lack of dry sand for sunbathers. The absence of formerly wide dry sandy beaches forces beachgoers to congregate on what dry sand there is at the pocket beaches. Photo by W. J. Elliott, 6-21-2012.*

Photo 19. *Concrete armor being erected to protect both lower and upper sea cliffs at the Sea Scape Chateau condominium project. Photo by W. J. Elliott, 12-10-1983.*

REFERENCES

Carter, G. F., 1957, Pleistocene Man at San Diego: Johns Hopkins Press, Baltimore, 400p.

Flick, R. E., 1993, The Myth and Reality of Southern California Beaches: Shore and Beach, Journal of the American Shore and Beach Preservation Association, v. 61, n. 3, July 1993, pp. 3–13.

Hanna, M. A., 1926, Geology of the La Jolla Quadrangle, California: University of California Publications, Bulletin of the Department of Geological Sciences, v. 16, n. 7, pp. 187–246, map scale 1" = 1 mile.

Kennedy, M. P., 1975, Geology of the San Diego Metropolitan Area, California, Section A, Western San Diego Metropolitan Area: California Division of Mines and Geology Bulletin 200, pp. 1–39, map scale 1" = 2,000'.

Nordstrom, C. E., and Inman, D. L., 1973, Beach and Cliff Erosion in San Diego County, California, in Ross, A., and Dowlen, R. J., eds., Studies on the Geology and Geologic Hazards of the Greater San Diego Area, California: San Diego Association of Geologists field trip guidebook, May 1973, pp. 125–131.

Strand, R. G., compiler, 1962, Geologic Map of California, San Diego - El Centro Sheet: California Division of Mines and Geology, map scale 1" = 4 miles.

Rogers, T. H., compiler, 1966, Geologic Map of California, Santa Ana Sheet: California Division of Mines and Geology, map scale 1" = 4 miles.

San Diego, County of, 1975, Ortho-Photo Topographic Map, No. 298-1683, scale 1" = 200', contour interval 5'.

San Diego, County of, 1985a, Ortho-Photo Topographic Map, No. 294-1683, scale 1" = 200', contour interval 5'.

San Diego, County of, 1985b, Ortho-Photo Topographic Map, No. 302-1683, scale 1" = 200', contour interval 5'.

United States Department of Agriculture, 1939, Stereographic black and white aerial photographs, Nos. AXN-204-51 and 52, flown 4-16-1939, approximate scale 1" ~ 0.32 mile.

United States Department of Agriculture, 1953, Stereographic black and white aerial photographs, Nos. AXN-8M- 79, 80, and 81, flown 4-11-1953, approximate scale 1" ~ 0.34 mile.

United States Geological Survey, 1967-75, Del Mar, California, 7½' topographic quadrangle map, scale 1" = 2,000', contour interval 20'.

United States Geological Survey, 1968-75, Encinitas, California, 7½' topographic quadrangle map, scale 1" = 2,000', contour interval 20'.

Weber, F. H., Jr., compiler, 1963, Geology and Mineral Resources of San Diego County, California: California Division of Mines and Geology County Report 3, 309p., Plate 1 map scale 1" = 2 miles.

Del Mar Shores, Solana Beach, September 23, 2010.
Public beach access by stairs from Del Mar Shores Terrace.
Credit: Copyright © 2002-2012 Kenneth & Gabrielle Adelman, California Coastal Records Project, www.californiacoastline.org.

Fletcher Cove, Solana Beach, May 15, 2012.
Photo: Woodrow L. Higdon, Geo-Tech Imagery.

A CASE FOR THE CLEAN SAND LAYER WITHIN THE BAY POINT FORMATION IN SOLANA BEACH

Walter F. Crampton, R.G.E., R.C.E.
TerraCosta Consulting Group, Inc.
3890 Murphy Canyon Road, Suite 200, San Diego, CA 92123
wcrampton@terracosta.com.

James Knowlton, R.C.E., C.E.G.
Geopacifica, Inc. 3060 Industry Street Suite 105, Oceanside, CA 92054
alluvium@sbcglobal.net

Gregory A. Spaulding, P.G., C.E.G., C.H.G.
TerraCosta Consulting Group, Inc.
3890 Murphy Canyon Road, Suite 200, San Diego, CA 92123
gspaulding@terracosta.com

Braven R. Smillie, P.G., C.E.G.
TerraCosta Consulting Group, Inc.
3890 Murphy Canyon Road, Suite 200, San Diego, CA 92123
bsmillie@terracosta.com

An understanding of the origin of the clean sand lens above the terrace/formation contact in Solana Beach starts with some general background on the coastal geomorphology of southern California and northern Baja California. Contemporary examples of southern California/northern Baja California coastal dune fields and why they exist today, along with the coastal environment that existed 120,000 years ago when the ancestral Solana Beach coastal dunes were formed, complement our understanding.

For coastal dunes to develop, two criteria are necessary: 1. sufficient riverine sediment supply to both feed and nourish the dune system well in excess of the longshore littoral transport that moves the transient beach sands within each littoral cell, and 2. sufficient coastal sheltering unique to an area typically causing a wave shadow in the lee of an upcoast headland.

Along California's coastline, the headlands are typically formed by local crustal deformation and conjugate strike-slip faults (wrench-faulting) associated with movement between the North American and Pacific Plates, in turn causing differential erosion of the bedrock exposed on the coastline. This deformation is responsible for the many hooked bays and seaward-stepping headlands that exist along California's coastline.

Figure 1 provides an overall map of southern California showing the many fault-controlled natural hooked bays associated with the northerly-moving Pacific Plate relative to the North American Plate. The main contemporary movement of these two tectonic plates is spread across the San Andreas fault system, with thousands of subsidiary faults, many dormant, contributing to the evolution of the current coastal morphology.

As shown on Figure 1, two contemporary, relatively significant dune fields exist within southern California and northern Baja California: Pismo Beach and Primo Tapia. Figures 2 and 3 show GoogleEarth™ image close-ups of these two dune systems, along with the significant rivers that feed the dunes, and the northerly headlands that provide some protection. The hooked bay, similar for both sites, is also important because the angle of the shore face relative to the incident wave field results in a reduction, if not reversal, of sediment transport within the small sub-cell, further contributing to the stability of the dune field. In both of these locations, significant river systems feed an abundant supply of sediment and the hook-shaped geometry of the shoreline provides both a protected lee and a local reduction/reversal in wave-driven littoral transport capacity to sustain these relatively stable dune fields.

Stepping back in time around 120,000 years ago, when the climate was warmer and sea level was 20 feet higher, the Solana Beach shoreline was generally where Cedros Avenue is today. Weather conditions were stormier with significant rainfall producing abundant sediment along the ancestral shoreline. This geologic interval pre-dates the deposition of the Bay Point Formation sediments. The coastal bluff at the time was comprised entirely of the 50-million year old Eocene-age Torrey Sandstone, as was the gently seaward-dipping bedrock shore platform beneath the transient beach sands that existed at the time. Although we do not know, we may assume that in the vicinity of Tabletop Reef today, an erosion-resistant, Eocene-age oyster bed likely existed within the 120,000-year old shore platform surface. The oyster bed reef, combined with the displacement along the Tabletop Reef fault (that today results in the 60-foot westerly offset in the sea cliffs adjacent the northern end of Solana Beach) likely resulted in a more pronounced headland than what exists today. In conjunction, a similar relict reef may have extended further offshore and possibly to the south, and this resultant coastal morphology protected much of the ancestral Solana Beach shoreline.

The justification for a more pronounced headland is the relatively short time frame of the stillstand that existed 120,000 years ago (see Figure 4), and the expected less magnitude of erosion of the resistant, northerly portion of the ancestral Solana Beach shoreline.

We suspect that the oyster bed, 120,000 years ago, was buried 25 feet below the shore platform and did not exist continuously through the entire stratigraphic section up to the older shore platform which is exposed in the present-day coastal bluff. The presence of 50-million year old relict oyster beds, both upcoast and downcoast of San Elijo Lagoon, suggest that over an estimated 200,000 years, similar oyster beds likely existed. Due to the presence of the 10-foot thick clean sand layer and overlying dune field, it may be reasonably postulated that a substantial headland, along with these substantial oyster beds, did exist in the area, and that additional wave sheltering may have existed along the majority of the ancestral Solana Beach shoreline.

The San Diego Association of Geologists published a guidebook in 1985 titled, "On the Manner of Deposition of the Eocene Strata in Northern San Diego County," edited by Dr. Patrick L. Abbott with San Diego State University[1]. The guidebook contains several excellent papers on North County coastal geology. One paper in particular, *Eocene Lithofacies Exposed in Sea Cliffs from Leucadia to Cardiff by the Sea, San Diego County*, by Randall L. Irwin, nicely describes the evolution of the Eocene sediments along Solana Beach and Encinitas, with several stratigraphic columns along the Encinitas coastline juxtaposed due to faulting within the Eocene sediments. Although no stratigraphic columns were compiled specifically of the Eocene sediments exposed in the sea cliffs of Solana Beach, the lithofacies presented by Mr. Irwin, are also considered to be representative of these Eocene sediments, both southerly and northerly of the Tabletop Reef fault.

---

1   San Diego Association of Geologists, April 13, 1985, On the Manner of Deposition of the Eocene Strata in Northern San Diego County, Guidebook Edited by Patrick L. Abbott.

These lithofacies reaffirm the likely presence of Eocene-age oyster beds higher up in the stratigraphic section similar to those that are exposed in Tabletop Reef today. As importantly, the plausible spacial distribution of these oyster beds is also indicated in the illustration on Figure 11 of Mr. Irwin's paper, reproduced herein as Figure 5.

About 120,000 years ago, a relatively wide transient sandy beach likely existed along the ancestral Solana Beach shoreline, more or less where Cedros Avenue is today. Over the course of a few thousand years, as sea level dropped over 200 feet, the shoreline changed dramatically and shifted about 2.4 miles to the west. As sea level receded from the high stillstand, the stormier weather and abundant rainfall continued to contribute significant sediment to the shoreline, and this sediment was deposited on the 120,000 year old wave-cut platform along the southern California shoreline. From the U.S.-Mexico border northerly into northern San Diego County, this depositional sequence is referred to as the Bay Point Formation. Although widespread and well exposed throughout San Diego County, the Bay Point Formation is limited to the relatively narrow wave-cut platform seaward of the 120,000-year old ancestral sea cliff. The time of deposition of the Bay Point Formation is also relatively narrowly defined, likely limited to no more than 5,000 years of deposition, resulting in an age ranging from about 120,000 years ago to about 115,000 years ago. Similarly formed terrace deposits are also recognized both north and south of San Diego County.

The Geology Department at Southern Illinois University Carbondale offers an excellent publication on Exercises in Active Tectonics[2], with Exercise 6 providing a nice overview on coastal terraces and the effect that tectonic deformation and changes in sea level have had on the California coastline. We have reproduced one of the figures from Exercise 6 (see Figure 6), as it provides a nice overview of coastal landforms and the marine terraces that make up coastal California.

The composition of the Bay Point Formation consists of both marine and non-marine sediments deposited onto the old wave-cut platform, with the marine sediments being deposited immediately after sea level started

---

2 N. Pinter, 1996, Exercises in Active Tectonics: An Introduction to Earthquakes and Tectonic Geomorphology, Exercise 6, Coastal Terraces, Sea Level, and Active Tectonics. Prentice Hall: Upper Saddle River, NJ.

to recede. The majority of the non-marine sediments are derived from the adjacent upland materials. Within both Encinitas and Solana Beach, much of the upland materials that were eroded to form the Bay Point Formation consisted of the relatively sandy Torrey Sandstone, therefore, the Bay Point Formation consists of a relatively clean, slightly-cemented, silty sand. In contrast, much of the upland sources for the Bay Point Formation elsewhere within San Diego County (specifically southerly to the US-Mexico border) typically consisted of more clayey materials, including a variety of Eocene-age claystones and siltstones. These more clayey materials, when eroded and redeposited onto the older wave-cut platform, resulted in a more clayey Bay Point Formation. It is because of the lateral variation in source materials that the composition of the Bay Point Formation varies along its relatively narrow coastal exposure, from relatively clayey deposits in the southern part of the County to, relatively cleaner sands in the northern part of San Diego County.

Given the preceding, and again assuming a relatively sheltered coastal environment along the ancestral Solana Beach coastline, it appears that a relatively clean, 10-foot thick, transient sand beach may have covered the 120,000-year old wave-cut platform. Most of the subsequent deposition within the Solana Beach area was wind-blown eolian deposits, creating relatively large dune fields. These ancestral dunes make up the majority of the sloping terrace deposits exposed in the upper coastal bluffs of Solana Beach today. Nearly the entire coastal bluff of Solana Beach has a 10-foot thick, relatively clean beach sand layer between elevation 25 and 35 feet, which is in turn overlain by lightly cemented dune deposits. Both the clean sand layer and dune deposits are extremely susceptible to erosion. Notably, neither the highly erosive clean beach sand nor the dune deposits are encountered in Encinitas to the north.

The Corps of Engineers' Encinitas and Solana Beach Shoreline Feasibility Study provides an excellent overview of the Pleistocene-age bluff-forming units that comprise Encinitas and Solana Beach[3]. We have taken the liberty of reproducing Section D-2.1.2. - Pleistocene-Age Bluff-Forming Units from the Corps report below. We have added the bold print in portions of the reprinted text for emphasis.

---

3   U.S. Army Corps of Engineers, January 2003, Encinitas and Solana Beach Shoreline Feasibility Study, San Diego County, California, Appendix D.

The sloping upper portion of the coastal bluffs is comprised of late-Pleistocene marine terrace deposits, including sediments from a variety of geologic environments. The marine terraces are a landform consisting of bench-like relatively flat areas adjacent to the coastal bluffs. In the Encinitas and Solana Beach areas, the sediments consist of moderately-consolidated, poorly-indurated, light reddish-brown, silty fine sands and clean sands that include both nearshore marine sediments, and beach and dune sands. The marine terrace deposits overlie a wave-cut abrasion platform, formed on the Eocene bedrock approximately 120,000 years ago when sea level was 20 feet higher (Lajoie and others, 1992). At that time, the sea was at a high eustatic level due to substantial melting of the ice caps during an interglacial period. Today, the abrasion platform ranges in elevation from approximately 17 feet near Batiquitos Lagoon, to approximately 70 feet at San Elijo State Beach, with the majority of the abrasion platform elevation along the Solana Beach coastline at or near 25 feet (MSL datum). The difference in elevation is a result of variable regional uplift associated with gentle tectonic folding during the past 120,000 years. Based on their location underlying the major marine terrace adjacent to the coast and overlying the abrasion platform, the sediments in the coastal bluff of the Encinitas/Solana Beach coast are correlated with the Bay Point Formation (approximately 120,000 years old).

The terrace deposits throughout virtually the entire study area are capped by an approximately 10-foot-thick, iron-oxide-cemented, residual clayey sand deposit. This upper Bay Point, erosion-resistant capping material, formed by the concentration of clayey weathering products, secondary oxides of iron and aluminum, and leached and re-precipitated salts, is the result of long exposure to the elements during a period of tropical to temperate climate.

Throughout much of Solana Beach, horizontally-bedded clean sand beach deposits exist within the lower part of this geologic unit. **Wherever these clean sands are exposed by a cliff failure, the bluff becomes unstable and susceptible to failure. Ongoing and progressive upper-bluff failures continue to this day along the northerly portion of the Solana Beach coastline.** Overlying the beach sands are thick sand dune deposits, which comprise much of the middle Bay Point Formation in this area and likely part of a dune field that overran the beach deposits after the sea retreated. **These clean relic beach sands and thick overlying dune** deposits do not appear to exist along the Encinitas shoreline, and, for that matter, have not been encountered in other Bay Point Formation exposures extending from the Point Loma Peninsula in central San Diego, up to the northerly limits of San Diego County. Logs of two representative test borings drilled through these relatively clean sand deposits in Solana Beach are included in Figures D-3A through D-3E. Along the Encinitas coast, the middle Bay Point Formation is divided into sections by ledge-forming units created by short term operation of the same processes that formed the resistant cap of the upper Bay Point. Each ledge forming unit represents a period when sedimentation was interrupted long enough for the weathering process to add some induration to the sediments. As a result, the tall sections of loose dune sand, which are so problematic for bluff stability in Solana Beach, are absent in most of Encinitas.

Figure 7, also reproduced from the U.S. Army Corps of Engineers' Encinitas and Solana Beach Shoreline Feasibility Study, shows a southerly perspective of Solana Beach taken in 2001, with Tabletop Reef in the foreground and the dramatic perspective of the northerly fault-controlled headland that extends some distance seaward from the otherwise relatively linear Solana Beach shoreline. The 120,000-year old shoreline is approximately one block east of Pacific Coast Highway, visible along the left margin of the photograph. The Tabletop Reef fault, responsible for the pronounced northerly headland, also predated the 120,000 year old stillstand and may have contributed to an even more dramatic hook-shaped bay, possibly with one or several nearshore oyster bed reefs contributing to a stable beach face and dune system. This system was fed by the San Dieguito River, Escondido Creek, and San Marcos Creek, all of which contributed sediment to the Solana Beach littoral system.

Illustrated on Figure 8 is the postulated 120,000-year old Solana Beach and Encinitas shoreline, along with some nearshore, high-relief oyster reefs contributing to the 120,000 year old relatively stable beach and dune system.

Recognizing the variability of material type within the Bay Point Formation, the City of Solana Beach's coastal terrace is somewhat unique within the County of San Diego, having the only exposed relatively uniform clean sand layer atop the 120,000-year old wave-cut terrace overlain by an eolian dune system, which, unfortunately, when undermined becomes rather unstable, triggering progressive upper-bluff failures that may take decades to re-equilibrate. Ongoing erosion that has more or less continued since the devastating 1997–98 El Niño storm season continues to undermine and further destabilize these relatively fragile upper terrace deposits and places the City's bluff-top residences at risk. With over 45 coastal bluff failures in Solana Beach during the past 13 years, numerous examples exist where the clean sand layer becomes exposed and progressive upper-bluff failures occur within a period of two to three years, threatening existing bluff-top improvements. It is this relatively unstable geologic environment that has necessitated shoreline stabilization along much of the City's coastline north of Fletcher Cove. The clean sand lens instability has prompted the City of Solana Beach to develop and recommend "Preferred Bluff Stabilization Measures."

The southerly part of the Solana Beach coastline continues to have a wider sand beach than the area north of Fletcher Cove, and this beach provides increased protection to the southerly portion of the City's shoreline. Along Solana Beach's 1.4 miles of coastline, it is apparent that the presence of the wider sand beach along the southerly part of the City's shoreline provides significant protection to the southerly portion of the City's coastal bluffs, while the absence of a protective sand beach northerly of Fletcher Cove, allows for more severe marine erosion and subsequent undermining and destabilizing of the upper-bluff terrace deposits. In the area north of Fletcher Cove, bluff-top structures become threatened, and the public that traverses this section of beach is at much greater risk, from ongoing bluff failures.

The clean sand layer exposed within the coastal bluffs in Solana Beach, typically between elevation 25 feet and 35 feet (MSL), cannot stand vertical and once exposed, tends to continually ravel and slough undermining the overlying slightly cemented dune sands triggering yet additional failures. **Coastal consultants have argued that, in order to stabilize these clean sands, seawalls must extend up to the top of the sands to essentially encapsulate and stabilize this relatively fragile lower section of the upper-bluff terrace deposits. Based on the alarming number of upper bluff failures in Solana Beach over the past decade, this does appear to be a problem and, once these clean sands are exposed, they need to be stabilized to stop yet additional and more severe upper-bluff failures.**

ILLUSTRATIONS

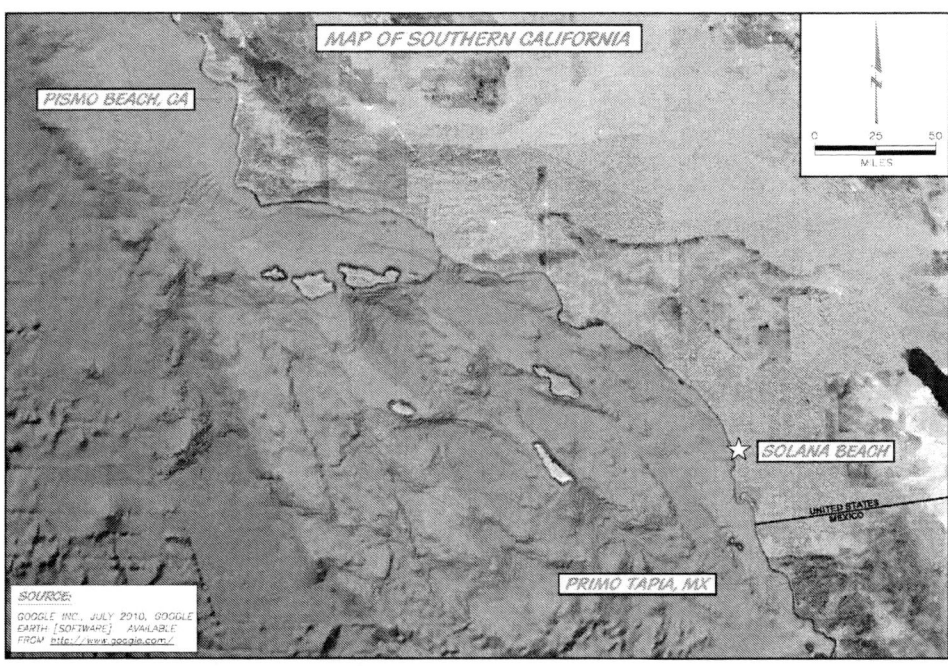

Figure 1. *Map of Southern California.*

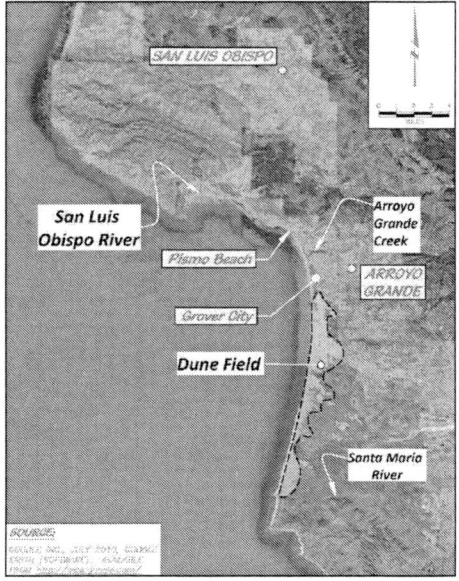

Figure 2. *Pismo Beach Dune Field.*

Figure 3. *Primo Tapia Dune Field.*

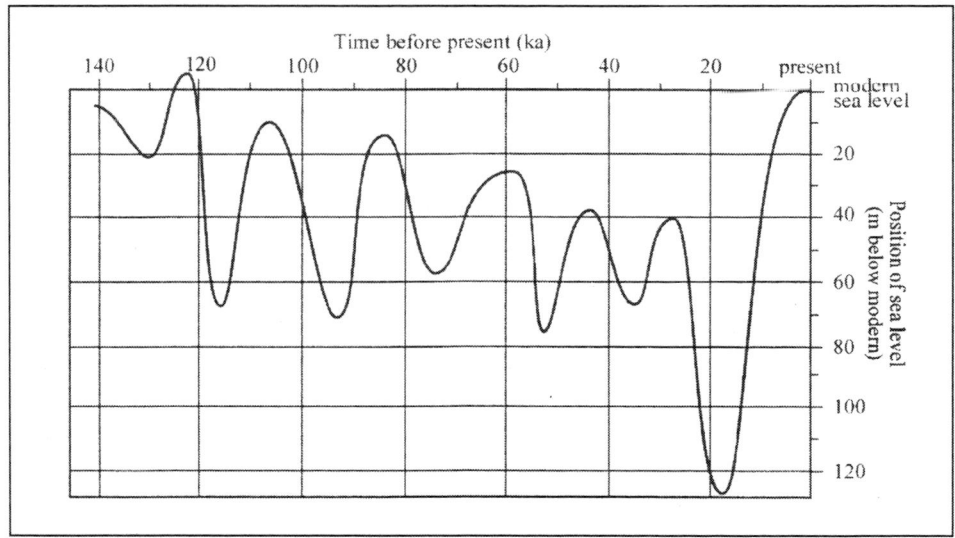

Figure 4. *History of fluctuating sea level during the last 140,000 years. Reference: Figure 6.3 of "Exercises in Active Tectonics: An Introduction to Earthquake and Tectonic Geomorphology, Exercise 6, Coastal Terraces, Sea Level, and Active Tectonics" by N. Pinter, dated 1996.*

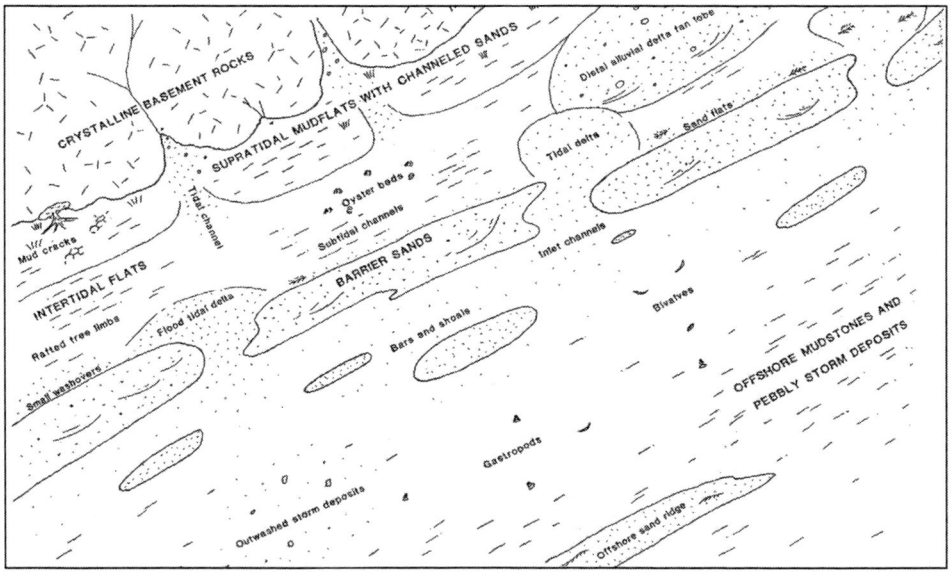

Figure 5. *Cartoon showing the theoretical spatial distribution of the depositional environments indicated by the various lithofacies recognized. Reference: Figure 11 of "On the Manner of Deposition of the Eocene Strata in Northern San Diego County," a San Diego Assoc. of Geologists Guidebook, edited by Patrick L. Abbott, dated April 13, 1985.*

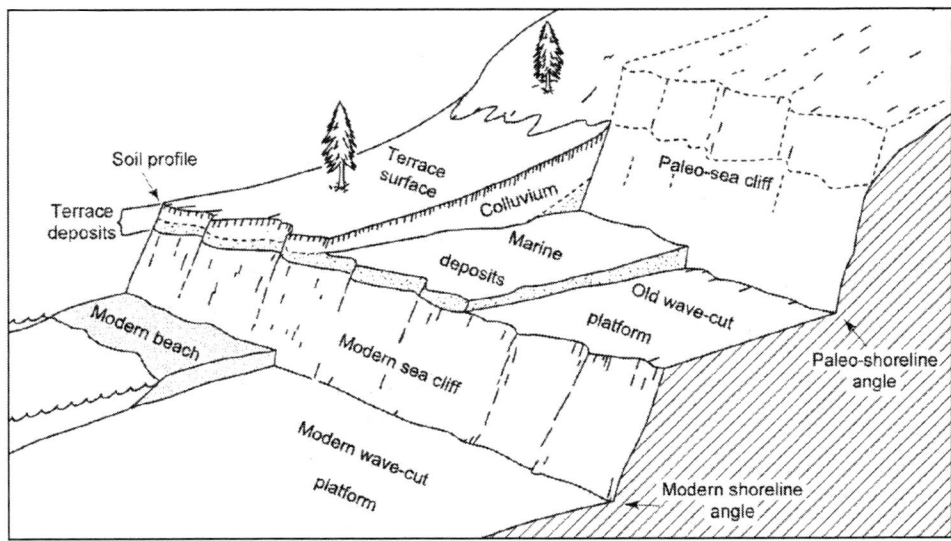

Figure 6. *An uplifted marine terrace and associated features (After Weber, 1983). Reference: Figure 6.4 of "Exercises in Active Tectonics: An Introduction to Earthquake and Tectonic Geomorphology, Exercise 6, Coastal Terraces, Sea Level, and Active Tectonics" by N. Pinter, dated 1996.*

Figure 7. *Figure 7: Solana Beach looking south. Tabletop Reef and fault-controlled headland in foreground. Aerial oblique photograph, 2001. U.S. Army Corps of Engineers.*

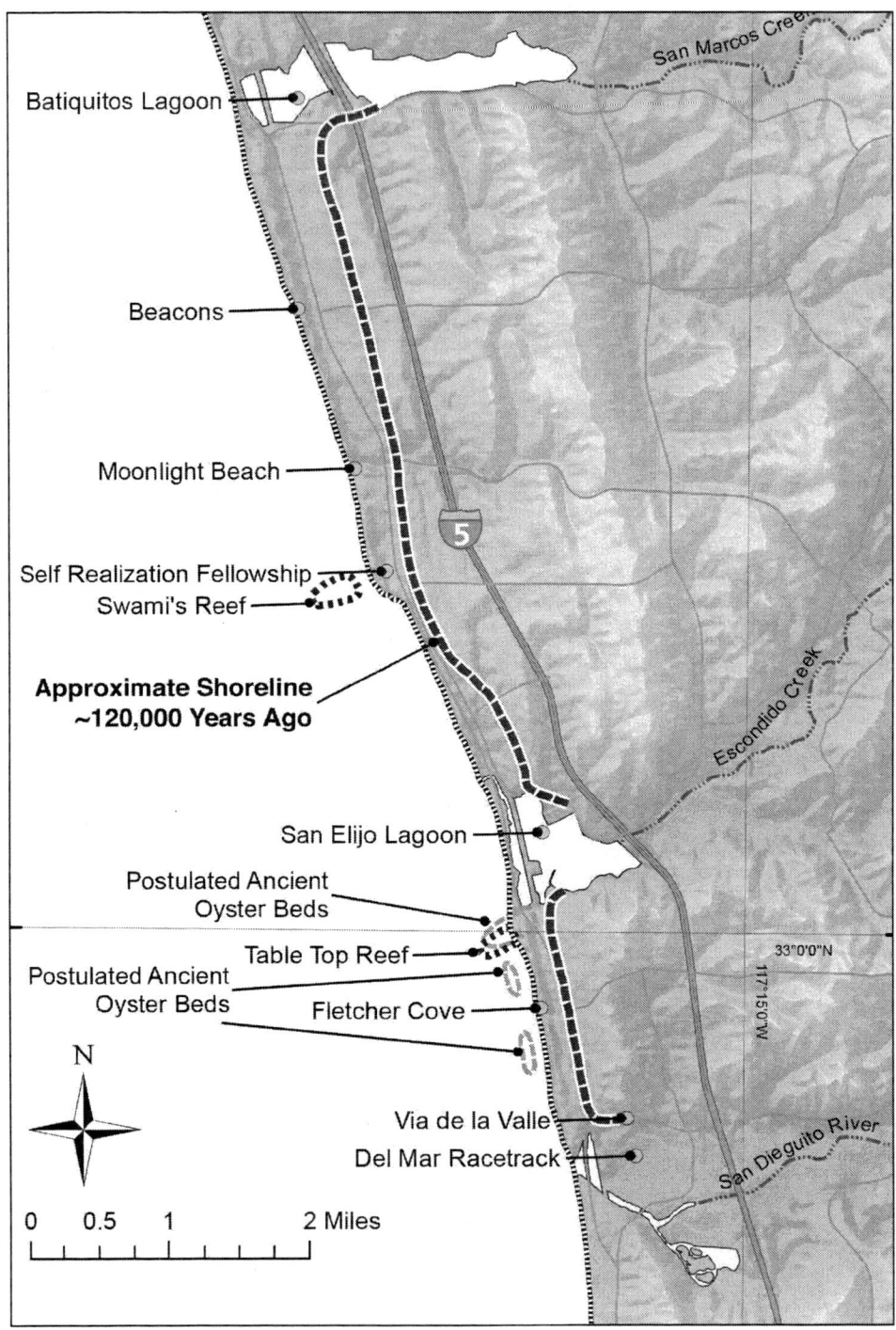

Figure 8. *Postulated 120,000 year old shoreline.*

Swami's Beach, Encinitas, October 24, 2004.
Credit: Copyright © 2002–2012 Kenneth & Gabrielle Adelman, California Coastal Records Project, www.californiacoastline.org.

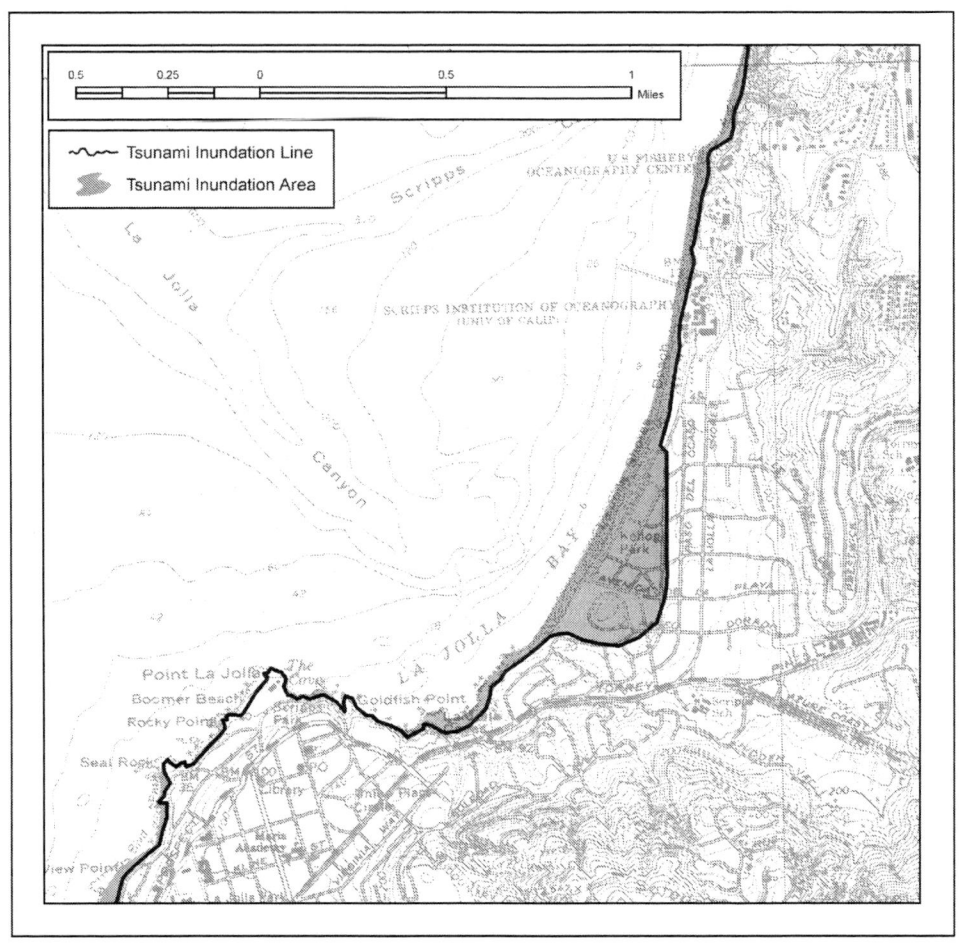

**Portion of La Jolla Quadrangle,
Tsunami Inundation Map for Emergency Planning, June 1, 2009.**
*(http://www.conservation.ca.gov/cgs/geologic_hazards/Tsunami
/Inundation_Maps/SanDiego/Pages/SanDiego.aspx)*

TSUNAMI HISTORY OF SAN DIEGO

Duncan Carr Agnew
Institute of Geophysics and Planetary Physics
Scripps Institution of Oceanography
University of California
La Jolla, CA 92093

The purpose of this paper is to list the available records of tsunamis at San Diego. For remote tsunamis the historical record is quite extensive; for local tsunamis data are essentially nonexistent.

A remote tsunami is one observed more than a few wavelengths from its point of generation. Large runup from remote tsunamis usually occurs when the offshore topography concentrates the tsunami energy. This does not seem to have happened at San Diego. Tide gauge records for San Diego bay extend from 1854 to 1872 and from 1906 to the present. In the 92 years of record, at least 19 tsunamis have been recorded. Most have been only a few tenths of a meter in height; for comparison, the diurnal range of tide at San Diego is 1.7 meters. The largest one was caused by the Chilean earthquake of May 1960. In San Diego it had a maximum range (peak-to-trough) of 1.5 meters, and produced strong currents which caused some damage to piers and which temporarily halted ferry service to Coronado.

In its recorded history (since the late 1700s) San Diego has experienced only one tsunami caused by a local earthquake. It was associated with the earthquake of May 27, 1862, which caused the most intense shaking known for San Diego (Legg and Agnew, 1979). Because the tide gauge was being repaired at the time, there is no quantitative record of the event, but an eyewitness account by the tidal observer, Andrew Cassidy, has been preserved[*]. At the time of the earthquake Cassidy was on the beach at La Playa, about 2 kilometers north of Ballast Point, on the east side of Point Loma. He wrote that, "The water in the Bay did not appear to be much agitated notwithstanding the sea run up on the beach between 3 and 4 feet, and immediately returned to its usual level." The value of 3 to 4 feet probably refers

---

[*] In a memorandum dated May 27, 1862, pasted on page 711 of the "Emigrant Notes" compiled by Benjamin Hayes (manuscript CE 62, Bancroft Library, Berkeley).

to the horizontal distance along the beach; the wave height would have been much less. Cassidy also noted falls of earth in the banks between La Playa and Point Loma. If one of these were large enough it might have caused a single wave in San Diego Bay similar to, but much smaller than, that caused by a rockslide in Lituya Bay, Alaska, in 1958 (Miller, 1960). In light of this possibility it would be premature to use this observation to conclude that tsunamigenic earthquakes can occur near San Diego. Such a conclusion, and any estimates of risk from local tsunamis, would have to come from a combined study of earthquake recurrence rates and types of faulting in the offshore area, together with model studies of tsunami generation, propagation, and runup. All the historical evidence shows is that damaging local tsunamis have not occurred at San Diego in the last two centuries.

The following list, which is based on that of Joy (1968), is limited to those tsunamis for which there is some evidence of a record at San Diego. For each date, the list gives the location and magnitude of the causative earthquake, followed by the size of the tsunami at tide gauges near San Diego. Where possible, earthquake magnitudes have been given according to the M_w scale introduced by Kanamori (1977). Three terms have been used to specify tsunami size. "Range" is the maximum value from peak to trough, also called "maximum rise or fall." "Height" (taken from Iida *et al.*, 1967) is the maximum positive departure from normal sea level. "Amplitude" is used when the definition of size in the original source is unclear. Though it is tempting to estimate a period for a tsunami record, it may not be too meaningful. Detailed analysis of the 1960 tsunami as recorded at La Jolla showed a broad spectrum (Miller *et al.*, 1962). Tsunami records from San Diego harbor give the general impression that the predominant periods are in the range of one-half to one hour.

Unless otherwise specified, the sources of information for this list are as follows. Earthquake locations before 1900 are from Iida *et al.* (1967); from 1900 to 1954 locations are from Gutenberg and Richter (1954), and M_s magnitudes are from Geller and Kanamori (1977) and Geller *et al.* (1978); after 1954 locations and M_s magnitudes are from epicenters lists published by the U. S. Coast and Geodetic Survey and its successor agencies. M_w magnitudes are either from Kanamori (1977) or have been computed from moment estimates. Tsunami information is from Iida *et al.* (1967). Other references are given in the individual listings.

TSUNAMIS AT SAN DIEGO 1854-2011

1854 July 24. No source is known, but the tidal observer at San Diego noted that on this date, "Water rose & fell nearly a foot in 10 minutes — currents set up also, harbor calm." (Andrew Cassidy, "Miscellaneous Notes on the Tide Gauge & Tidal Observations at San Diego -Cal: 1853-1854". Cassidy Papers, Serra Museum Library, San Diego).

1854 December 23. Japan, 34° N, 138° E. Range 0.1 m at San Diego (Bache 1855).

1856 August 23. Japan, 42° N, 141° E. Recorded on the coast of California (Joy, 1968).

1862 May 27. Earthquake at San Diego caused a small tsunami in San Diego Bay. See text for details.

1868 April 2. Hawaii, 19.3° N, 155.3° W (Wood, 1914). Height 0.1 m at San Diego.

1868 August 13. Chile, 18.5° S, 71° W. Amplitude 0.8 m at San Diego (Hilgard,1869).

1872 August 23. Davidson (1872) said that on this date a tsunami was recorded at San Diego, San Francisco, and Astoria. He used relative arrival times to infer a source in the northwest Pacific.

1906 January 31. Off the coast of Ecuador, 1° N, 81.5° W. M_w = 8.8. Recorded at San Diego.

1917 May 2. Kermadec Islands, 29° S, 177° W. M_s = 7.9. Recorded on the west coast of the U. S. (Heck, 1947).

1917 June 25. Tonga, 15.5° S, 173° W. M_s = 8.4. Recorded on the west coast of the U. S. (Heck, 1947).

1919 April 30. Tonga, 19° S, 172.5° W. M_s = 8.2. Recorded in California (Heck, 1947).

1922 November 10. Central Chile, 28.5° S, 70° W. M_w = 8.5. Height 0.2 m at San Diego.

1923 February 4. East coast of Kamchatka, 54° N, 161° E. M_w = 8.3. Height 0.2 m at San Diego.

1923 April 14. East coast of Kamchatka, 56.5° N, 162.5° E. M = 7.2 (Gutenberg and Richter, 1954). Height 0.1 m at San Diego.

1921 November 4. Off Point Arguello, California, 34.5° N, 121° W. M_w = 7.3 (Hanks et al., 1975). Range 0.006 m at La Jolla (Byerly, 1930).

1933 March 2. East of Honshu, 39.2° N, 144.5° E. M_w = 8.4. Height less than 0.1 m at La Jolla.

1944 December 7. Near Honshu, 33.7° N, 136° E. M_w = 8.1. Height 0.1 m at San Diego.

1946 April 1. Southern Alaska, 52.75° N, 163.5° W. M_w = 8.4 (Kanamori, 1972). Range 0.43 m at La Jolla, 0.37 m at San Diego (Green, 1946; Symons and Zetler, 1960).

1952 March 4. Hokkaido, Japan, 42.5° N, 143° E. M_w = 8.1. Range 0.02 m at La Jolla (Munk, 1953).

1952 November 5. Off east coast of Kamchatka, 52.7° N, 159.5° E. M_w = 9.0. Range 0.24 m at La Jolla, 0.7 m at San Diego (Zerbe, 1953).

1957 March 9. Rat Islands, 51.3° N, 175.8° W. M_w = 9.1. Range 0.6 m at La Jolla, 0.45 m at San Diego (Salsman, 1959).

1960 May 22. Coast of central Chile, 39.5° S, 74.5° W. M_w = 9.5. Range 1 m at La Jolla, 1.5 m at San Diego (Symons and Zetler, 1960; Miller et al., 1962). Some damage to piers and moorings in San Diego Bay.

1964 March 27. Southern Alaska, 61° N, 141.8° W. M_w = 9.2. Range 0.7 m at La Jolla, 1.1 m at San Diego (Spaeth and Berkman, 1964).

1968 May 15. East of Honshu, 29.9° N, 129.4° E. M_w = 8.2. Amplitude 0.1 m at La Jolla (Joy, 1968).

1975 November 29. Hawaii, 19.3° N, 155° W. M_s = 7.1. Amplitude 0.3 m at La Jolla, 0.12 m at San Diego, 0.37 m at Imperial Beach (Spaeth, 1976).

1977 June 22. Tonga Islands region, 22.9° S, 175.9° W. M_s = 7.2. Height 0.080 m at San Diego.*

1994 October 4. Kuril Islands, Russia, 43.8° N, 147.321. M_s = 8.3. Height 0.030 m at La Jolla.*

1995 July 30. Near coast of northern Chile, 23.3° S, 703° W. M_s = 8.0. Height 0.050 m at San Diego.*

2001 June 23. Near coast of Peru, 16.3° S, 73.6° W. M_s = 8.4. Height 0.050 m at La Jolla, 0.050 m at San Diego.*

2004 December 26. Off west coast of northern Sumatra, Indonesia, 3.3° N, 95.9° E. M_s = 9.1. Height 0.060 m at La Jolla, 0.160 m at San Diego.*

2006 May 3. Tonga Islands, 20.2° S, 174.1° W. M_s = 8.0. Height 0.040 m at La Jolla.

2006 November 15. Kuril Islands, Russia, 46.6° N, 153.3° E. M_s = 8.3. Height 0.100 m at La Jolla, 0.090 m at San Diego.*

2007 April 1. Bougainville, Solomon Islands region, 8.5° S, 157° E. M_s = 8.1. Height 0.100 m at San Diego.*

2007 August 15. Near coast of Peru, 13.4° S, 76.6° W. M_s = 8.0. Height 0.050 m at San Diego.*

2010 February 27. Near coast of central Chile, 36.1° S, 72.9° W. M_s = 8.8. Height 0.600 m at La Jolla, 0.400 m at San Diego.*

2011 March 11. Near east coast of eastern Honshu, Japan, 38.3° N, 142.4° E. M_s = 9.0. Heights 0.390 m and 0.900 m at La Jolla, 0.630 m at San Diego.*

* The author provided updated historical tsunami data 1977–2011 (NOAA, 2012).

ACKNOWLEDGEMENTS

I should like to thank B. D. Zetler for help and comments.

REFERENCES

Bache, A. D., 1855, Notice of earthquake waves on the western coast of the United States, on the 23d and 25th December, 1854: U. S. Coast Survey Annual Report, 1855, p. 342–346.

Byerly, P., 1930, The California earthquake of November 4, 1927: Seismological Society of America Bulletin, v. 20, p. 53–66.

Davidson, G., 1872, Remarks on recent earthquake waves: California Academy of Sciences Proceedings, ser. 1, v. 4, p. 268.

Geller, R. J., and Kanamori, H., 1977, Magnitudes of great shallow earthquakes from 1904 to 1952: Seismological Society of America Bulletin, v. 67, p. 587–598.

Geller, R. J., Kanamori, H., and Abe, K., 1978, Addenda and corrections to "Magnitudes of great shallow earthquakes from 1904 to 1952": Seismological Society of America Bulletin, v. 68, p. 1763–1764.

Green, C. K., 1946, Seismic sea wave of April 1, 1946, as recorded on tide gages: American Geophysical Union Transactions, v. 27, p. 490–500.

Gutenberg, B., and Richter, C. F., 1954, Seismicity of the Earth and associated phenomena, Princeton, Princeton University Press, 310 pp.

Hanks, T. C., Hileman, J. A., and Thatcher, W., 1975, Seismic moments of the larger earthquakes of the southern California region: Geological Society of America Bulletin, v. 86, p. 1131–1139.

Heck, N. H., 1947, List of seismic sea waves: Seismological Society of America Bulletin, V. 37, p. 269–286.

Hilgard, J. E., 1869, The earthquake wave of August 14, 1868: U. S. Coast Survey Annual Report, 1869, p. 233.

Iida, K., Cox, D. C., and Pararas-Carayannis, G., 1967, Preliminary catalog of tsunamis occurring in the Pacific Ocean: Hawaii Institute of Geophysics Data Report 5 (HIG 67-10).

Joy, J. W., 1968, Tsunamis and their occurrence along the San Diego County coast: Report to the Unified San Diego County Civil Defense and Disaster Organization.

Kanamori, H., 1972, The mechanism of tsunami earthquakes: Physics of the Earth and Planetary Interiors, v. 6, p. 346–359.

Kanamori, H., 1977, The energy release in great earthquakes: Journal of Geophysical Research, v. 82, p. 2981-2987.

Legg, M. R., and Kennedy, M. P.; 1979, Faulting Offshore San Diego and Northern Baja California, in Earthquakes and Other Perils, San Diego Region. Patrick L. Abbott and William J. Elliott, eds. San Diego Association of Geologists, November.

Miller, D. J., 1960, Giant wave in Lituya Bay: Seismological Society of America Bulletin, v. 50, p. 253-266.

Miller, G. R., Munk, W. H., and Snodgrass, F. E., 1962, Long-period waves over California's continental borderland. Part II. Tsunamis: Journal of Marine Research, v. 20, p. 31-41.

Munk, W. H., 1953, Small tsunami waves reaching California from the Japanese earthquake of March 4, 1952: Seismological Society of America Bulletin, v. 43, p. 219-222.

National Oceanographic and Atmospheric Administration (NOAA), 2012, National Geophysical Data Center, Global Historical Tsunami Database. http://www.ngdc.noaa.gov/hazard/tsu_db.shtml

Salsman, G., 1959, The tsunami of March 9, 1957, as recorded at tide stations: U. S. Coast and Geodetic Survey Technical Bulletin 6.

Spaeth, M., 1976, Tsunamis: United States Earthquakes 1975, p. 115-116.

Spaeth, M., and Berkman, S. C., 1964, The tsunami of March 28, 1964, as recorded at tide stations: U. S. Coast and Geodetic Survey Technical Report 6.

Symons, J., and Zetler, B. D., 1960, The tsunami of May 22, 1960, as recorded at tide stations: U. S. Coast and Geodetic Survey Preliminary Report.

Zerbe, W. B., 1953, The tsunami of November 4, 1952, as recorded at tide stations: U. S. Coast and Geodetic Survey Special Publication 300.

Updated in 2012 by contribution by the author.
Modified from: Earthquakes and Other Perils, San Diego Region. Patrick L. Abbott and William J. Elliott, eds. San Diego Association of Geologists, November 1979.

THEORETICAL ASPECTS OF TSUNAMIS ALONG THE SAN DIEGO COASTLINE

W. G. Van Dorn
Ocean Research Division
Scripps Institution of Oceanography
La Jolla, CA 92093

If one were asked to pick the safest place in the Pacific Margin from the standpoint of geologic hazards from tsunamis, San Diego County would be a top contender. Of the five greatest tsunamis occurring in this century, not one of them produced effects readily apparent even to a skilled observer walking along the seacoast (Table 1). The greatest single excursion of sea level (1.0 m) was recorded at La Jolla for the tsunami of May 22, 1960, which was produced by the largest earthquake ever recorded (Richter Magnitude 8.5). The strong alternating currents engendered with in San Diego Bay by this same tsunami ripped loose a few hundred feet of rotted wharfage, and caused the Coronado Ferry to discontinue service for several hours for the first time in 80 years of operation.

Table 1. *Maximum Recorded Local Tide Gauge Excursions Δh (cm) for Five Major Tsunamis (USC&GS Reports)*

	Tsunami Date	4-1-46	11-2-52	7-1-57	5-22-60	3-28-64
	Source Location	Aleutians	Kamchatka	Aleutians	Chile	Alaska
Δh	San Diego (Broadway Pier)	37	70	46	137	110
	La Jolla (SIO Pier)	43	24	61	100	76

Because the above tsunamis have originated from all representative directions where tectonically active tsunamigenic sources are thought to exist around the Pacific, it is considered unlikely that new surprises may alter present confidence in San Diego's immunity from tsunami damage. The physical reasons underlying this confidence are, however, more associative than theoretical.

(1) It has long been recognized that major tsunamis originate from vertical

dislocations of large crustal blocks (105 km^2) by a meter or so, and that such dislocations are primarily confined to the Pacific trench system and have recurrence intervals of several hundreds of years.

(2) Recent numerical calculations of the resulting patterns of water waves suggest that most large tsunamis are very similar in deep water, and that their local intensity depends mainly on source orientation, distance, and angle of approach to a remote point.

(3) Nature has providentially arranged matters so that waves from currently active foci approach southern California from nearly glancing incidence; further protection is afforded by wave transformation and/or reflection at the margin of our broad coastal shelf.

Thus, calculations of regional effects serve only to confirm historical precedent as regards remotely generated tsunamis. As regards local generation, the prognosis is similarly optimistic, but for different reasons.

(1) Relative motion of the Pacific crustal plate, vis-à-vis the continental plate off California, is principally horizontal. Hence, local earthquakes along the San Andreas fault system are prone to strike-slip motions. This circumstance is the strongest argument against the occurrence of a major tsunamigenic earthquake in this sector of the Pacific.

(2) Relatively minor local tsunamis have been produced by small earthquakes on the continental shelf north of Point Fermin, and there is at least a verbal description of a three-foot wave in San Diego Bay associated with the earthquake of May 27, 1862, the only positive evidence for significant thrusting of any magnitude is the possible connection of an offshore escarpment with the Agua Blanca fault (Legg and Kennedy, 1979).

(3) Given the dimensions of the above fault zone, one could assume a maximum credible areal dislocation, and use present numerical methods to estimate the magnitude of shoreline effects along the San Diego coastline.

So far, this has not been done. Thus, one can conjecture that (1) it would be highly unlikely that a destructive tsunami could be produced by any reasonable dislocation (1 m) along this fault zone, and (2) that the most likely result would be minor flooding of the low-lying areas in Imperial Beach and Mission Beach, if it happened to coincide with a high tide.

REFERENCE

Legg, Mark R., and Michael P. Kennedy; 1979, Faulting Offshore San Diego and Northern Baja California, in Earthquakes and Other Perils, San Diego Region. Patrick L. Abbott and William J. Elliott, eds. San Diego Association of Geologists, November.

Adapted from: Earthquakes and Other Perils, San Diego Region. *Patrick L. Abbott and William J. Elliott, eds. San Diego Association of Geologists, November 1979.*

Rose Canyon, San Diego, November 3, 1990.
Photo: Woodrow L. Higdon, Geo-Tech Imagery.

THE ROSE CANYON FAULT ZONE IN SAN DIEGO

Thomas Rockwell
Earth Consultants International
1642 East Fourth Street
Santa Ana, CA 92701-5148

ABSTRACT

The Rose Canyon fault bisects the City of San Diego, producing much of the unique beauty of the city with the uplift of Mt. Soledad and subsidence producing the natural harbor of San Diego Bay. Geologic studies demonstrate that the late Quaternary slip rate is in the range of 1–2 mm/yr, which although only a fraction of the plate margin slip budget, has the potential to produce major damage in "America's finest city." Paleoseismic trenching in La Jolla and downtown San Diego indicate that the most recent surface rupture occurred only a few hundred years ago, sometime after about AD 1523 but prior to the establishment of the San Diego mission in 1769. Displacement in this earthquake may have been as much as 3 m based on 3-dimensional trenching. Using this displacement and slip rate, the average return period should be on the order of 1500–3000 years, suggesting that San Diego may be safe for the near future. However, limited observations suggest that the Rose Canyon fault behaves in a clustered mode, where earthquakes are clustered in time, rather than in a quasi-periodic fashion. If correct, and considering that the rupture in the past few hundred years appears to have been the first large earthquake in more than five thousand years, San Diego may have recently entered a renewed period of activity.

INTRODUCTION

The Rose Canyon fault is the primary potential seismic hazard to the City of San Diego due to the fault's proximity: a moderately large earthquake could potentially do significant damage to the City and surroundings, both in terms of shaking and ground rupture. In this paper, I briefly describe the current state of knowledge on the location, slip rate, timing and size of past large events on the fault in the City, and speculate on its behavior and the potential implications for the likelihood of future damaging earthquakes.

In the broad scheme of plate tectonics, the Rose Canyon fault is a bit player, as it accommodates less than 5% of the total plate motion (Figure 1). The fault is part of a coastal system of faults that accommodate 6–7 mm/yr between San Clemente Island and the mainland. From the south, the Agua Blanca fault in northern Baja California feeds slip northward at about 5–6 mm/yr (Rockwell et al., 1993), with this deformation principally accommodated by the Rose Canyon, Coronado Bank, and San Diego Trough faults. Farther offshore, the San Clemente fault appears to extend southward of the Agua Blanca fault (Legg, 1985), and may be part of a zone that extends the length of the Baja peninsula. To the north, the Rose Canyon fault becomes the Newport-Inglewood fault, and this zone of active faulting extends northwestward through the Los Angeles basin to the Transverse Ranges (Figure 1).

Within San Diego, the Rose Canyon fault zone comprises a broad zone of active and inactive strike-slip and normal faults (Figure 2). Mt. Soledad is interpreted as a pressure ridge where the fault makes a left or transpressive bend, producing local uplift. Kern and Rockwell (1992) show that this deformation has continued into the late Quaternary with uplift of Pleistocene marine terraces. Similarly, San Diego Bay represents extension across a large right (releasing) step-over, where the majority of the Rose Canyon slip steps westward to the Descanso fault off northern Baja California (Figure 2). A consequence of this step is a minor component of normal faulting along the La Nacion fault zone, located east of the dextral fault zone, as well as transtensive faulting in San Diego Bay itself (see Gingery et al., 2010). The La Nacion fault is only expressed in the geology and topography adjacent to where the Rose Canyon fault bends south and steps to the Descanso fault. Although not proven definitively active, the La Nacion fault is part of the Rose Canyon deformation and, by nature of its sense of slip and structural ties to the Rose Canyon fault, likely continues to accommodate minor extension associated with the San Diego Bay step-over. One possible reason that geologists have not found definitive proof of its Holocene activity is that the discrete motion on the La Nacion fault is expected to be small on an event by event basis, so its expression in the active soil could easily be obscured.

The Rose Canyon fault is believed to have begun motion in the late Pliocene (Ehlig, 1980), although absolute constraints place it as post-Eocene and pre-late Pliocene based on the observation that the Upper Pliocene San Diego Formation appears less deformed than the Eocene rocks (Kennedy et al., 1975b). Ehlig (1980) interpreted the Pliocene Embayment as a direct consequence of the initiation of the Rose Canyon fault, and it is into this embayment that the San Diego Formation accumulated. If correct, at least the upper member of the San Diego Formation is a tectono-stratigraphic unit that potentially records the history of motion of this fault.

The San Diego Formation is generally considered Pliocene in age, and analysis of foraminifera and calcareous nannoplankton mixed in with the classic molluscan index fossil *Patinopecten healeyi* indicate an age of 3.8-4.2 Ma for the lower member (Boettcher, 2001; Kling, 2001). The upper member has been suggested to be as young as early Pleistocene (Deméré, 1982). Thus, strata of the San Diego Formation may span several million years in age, which is expected if part of the strata accumulated in an active tectonic setting.

Total displacement on the Rose Canyon fault is not well-resolved, but Kies (1982) suggests that facies within the Eocene Mount Soledad Formation are offset about 4 km in a right-lateral sense. Similarly, Moore and Kennedy (1975) and Kennedy (1975) suggest several kilometers of right-lateral displacement based on the observation that the Pliocene San Diego Formation is found several kilometers farther north on the west side of the fault. They suggest that as much as 4-6 km of displacement has occurred on the northern margin of this Pliocene embayment.

Combining the 4–6 km estimates of displacement with the ~4 Ma maximum age of the San Diego Formation yields a minimum lifetime slip rate for the fault of about 1 mm/yr. This minimum rate is in agreement with the minimum Holocene rate, as discussed below. A more reasonable lifetime rate is to accommodate the 4-6 km of displacement since the late Pliocene, which yields a slip rate closer to 2 mm/yr. In any case, the current rate is more significant in terms of local hazard estimates, as discussed below.

GEOMORPHIC EXPRESSION

The Rose Canyon fault is well-expressed in the landscape of San Diego. The transpressive bend at La Jolla has produced uplift of Mt. Soledad, and the transtensive step has down-dropped the area of San Diego Bay. In fact, without the Rose Canyon fault, San Diego would never have had the World-class natural harbor that led to its development into a World-class city.

On a finer scale, Lindvall and Rockwell (1995) analyzed early aerial photography from the 1920s to 1940s that predate the major expansion of the City after World War II. They found abundant evidence of youthful faulting between La Jolla and the San Diego River in the form of deflected and offset stream channels, small pressure ridges, a sag depression, and a scarp across the Holocene terrace to Rose Creek. Figure 3 shows a few of these features in the area near their trench, and Figure 4 shows deflected drainages and scarps interpreted from 1928 aerial photography in the Old Town area. Based on its geomorphic expression alone, the Rose Canyon fault is an active strike-slip fault.

SLIP RATE

The geologic late Quaternary slip rate of the Rose Canyon fault is derived from both short and long-term measurements. Lindvall and Rockwell (1995) excavated twelve trenches in a 3-dimensional excavation on a low terrace to Rose Creek and exposed a small gravel-filled channel that was radiocarbon dated to younger than 8.1 ka (calibrated years B.P.) The lateral excavations exposed remnants of the displaced channel across several strands of the fault in a 3 m-wide zone, but the unit containing the gravel was cut out by grading west of the fault (the site was graded and buried by up to 2.5 m of mechanically-emplaced fill in 1960). Consequently, Lindvall and Rockwell (1995) resolved only a minimum of 8.7 m of right-lateral displacement for the past 8.1 ka (maximum age), yielding a minimum Holocene slip rate of about 1.1 mm/yr. Furthermore, their study was conducted on the Mt. Soledad strand of the Rose Canyon fault zone, so this is a gross minimum for the entire zone if other strands are also active. Indeed, the Country Club strand is seen as a lineament crossing the Holocene terrace to Rose Creek in 1928 photography (Figure 5), and the Rose Canyon strand has not been trenched at this fault latitude. Considering this, Lindvall and Rockwell (1995) estimated the actual rate at 1.5±0.5 mm/yr, but with the upper limit on the rate being poorly constrained.

The stream deflections in Old Town, as shown in Figure 4, suggest a rate that is towards the high end of that suggested by Lindvall and Rockwell (1995). Several parallel channels are deflected at the fault by about 250±50 m. The channels incise a broad terrace surface which, based on its elevation of 12–18 m, is likely the Nestor terrace equivalent (Kern and Rockwell, 1992). The Nestor terrace is well dated in San Diego at about 120 ka (Ku and Kern, 1974) and corresponds to the global highstand during oxygen isotope stage 5e (using the nomenclature of Shackleton and Opdyke, 1973). If the stream deflections accurately reflect lateral displacement along the fault, which is a reasonable interpretation, this resolves to a longer-term slip rate of about 2±0.4 mm/yr. This rate should be valid for the entire fault zone, as the streams cross the entire fault zone and the fault only splays into multiple strands south of Old Town.

TIMING AND SLIP IN PAST EVENTS

After the work of Lindvall and Rockwell (1995), quite a bit of new information has come to light on the Holocene history of faulting. Two sites, one in downtown San Diego and one in La Jolla, have yielded radiocarbon data that indicate that the most recent surface rupture on the fault has occurred in the past few hundred years (Rockwell and Murbach, 1996; Grant and Rockwell, 2002). Moreover, information collected during the 3-dimensional trenching study includes observations that relate to both timing and magnitude of past displacements. As the Lindvall and Rockwell (1995) paper focused primarily on the slip rate, one of the logs is presented here (Figure 6) and reinterpreted in relation to other useful information on timing and slip.

Figure 6 shows the log from the trench T4 exposure from the Lindvall and Rockwell's (1995) study. The channel that was used to resolve a minimum of 8.7 m of lateral displacement (Figure 7) is embedded in units C1 and C2 (Figure 6), but the actual top of the channel fill is near the top of unit C1. Thus, the age of the channel is best defined by the age of upper unit C1. This stratum did not yield any charcoal for radiocarbon dating, but the underlying unit C2 contained charcoal that yielded an age of ~8.6 ka (calibrated). However, unit C3 yielded a younger age of ~8.13 ka (calibrated), indicating that units C2 and C1 are younger than 8.1 ka. Lindvall and Rockwell (1995) argue that the actual age of the channel is likely close to 8.1 ka because all of the dates from units C2 and C3 are similar and because the soil separating these units is very weakly developed and cannot represent much time. Nevertheless, this date is still a maximum age for the units that host the gravel channel, so the resulting slip rate is a minimum value.

The soil that caps the deposits is moderately developed with an argillic horizon (Table 1, profiles RCF 1, RCF 2 and RCF 3, all of which were described in trench exposures on the northeast side of the fault). The alluvium into which the soil is developed contains abundant fines (silt and clay), so argillic horizons will form fairly rapidly in the coastal environment of southern California due to the presence of abundant sodium (Rockwell et al., 1985; Rockwell, 2000). Nevertheless, this argillic horizon had common thin to moderately thick clay films and moderately-developed structure, which represents significant development requiring many thousands of years of time for its formation. Comparison of the soil development index (SDI)(Harden, 1982) for these profiles to dated soils elsewhere in coastal southern California (Rockwell et al., 1985; and unpublished data) suggests 5–10 ka for the age of this soil, although the radiocarbon dates from the underlying stratigraphy require the soil to be less than 8 ka. For the purposes of this study, a minimum age of about 5 ka is assigned to the surface soil capping the faulted stratigraphy at Rose Creek. This point is important in understanding the timing of ruptures along the fault.

The log of trench T4 also contains direct evidence of several surface ruptures during the period of deposition of the section. Several fault strands (strands 1, 2, 3 and the western strand of fault 4, as indicated on Figure 6) all appear to have ruptured during an event that occurred between deposition of units C3 and C2, as indicated by the black stars in Figure 6. These fault strands have splays that are terminated and capped by unfaulted stratigraphy, and there is alluvium or colluvium of unit C3 that is truncated at fault strand 1. As the age of unit C3 has been determined to be about 8.1 ka by radiocarbon dating (Lindvall and Rockwell, 1995), this event must have occurred at or very soon after that time.

An earlier event during deposition of unit C4 is suggested by the dramatic increase in thickness of unit C4 across fault strand 4: this relationship was observed in many exposures. The massive (colluvial) character of unit C4 east of the fault, along with the observation that the thickness of unit C4 tapers away from fault strand 4, support the idea that an event with some vertical displacement occurred during deposition of unit C4, and that unit C4 is, in part, a colluvial wedge of material shed from the scarp. If correct, this event occurred between deposition of units D and the top of C4, both of which yielded radiocarbon ages of about 9.3 ka (calibrated), so an age of ~9.3 ka is assigned for this event.

Unit C2 is degraded and observed to thin across the scarp between fault strands 1 and 4 (hollow star in figure 6), and unit C1 has finely bedded strata that appear to have been deposited in angular unconformity against a scarp between fault strands 2 and 3. As unit C2 also contains finely-bedded strata, this unit cannot be interpreted as a colluvial deposit, and the thinning therefore most likely represents erosion across the scarp after slip occurred on the fault. Thus, a surface-rupturing event that produced some vertical displacement is interpreted to have occurred after deposition of unit C2 and before deposition of unit C1.

These observations argue that as many as three surface ruptures may have occurred on the fault during deposition of units C4 through the base of C1, and all of these apparently occurred in the early Holocene in a fairly narrow window of time. Further, as the channel that was used to resolve slip is embedded in unit C1, none of these events contributed to the 8.7 m of displacement documented by Lindvall and Rockwell (1995).

Fault strands 2, 3 and 4 all have elements that die upward into unit C1 and do not displace the capping soil, including the well-formed Bt horizon, as indicated by the gray stars in figure 6. This observation requires one or more surface ruptures to have occurred between deposition of unit C1 (and the gravel channel) and the development of the soil, and these ruptures produced offset of the gravel channel embedded near the top of unit C1. In figure 7, these fault strands produced at least 5.7 m of cumulative lateral displacement, based on offset of the channel. The timing of these displacements are poorly constrained, but the fact that the soil is developed across these strands and is not offset by them indicates that none of these faults have likely moved in many thousands of years, and possibly not since soon after deposition of the section and the onset of soil development.

In contrast, the eastern two strands of fault 4 sharply offset the surface soil, including the topsoil (A) horizon. Reconstructing the vertical separation of the Bt, C1 and C2 horizon suggest that all are vertically displaced by the same amount, and this relationship was observed in many exposures during the 3D trenching. This, in turn, argues that a single rupture produced this deformation because multiple events should have sustained erosion and deposition of colluvium between ruptures unless there was essentially no time to allow for such erosion. As erosion of a scarp can begin within the first winter after a scarp-forming event, and as there is no evidence for such an event, it is likely that this displacement represents only a single slip surface rupture.

This has important implications from two perspectives. First, the most recent rupture appears to follow a rather lengthy period of time, as significant time is required to have allowed the soil to develop to its current state. After soil development, it appears that only the most recent event has produced slip on the Mt. Soledad strand of the Rose Canyon fault zone, which is interpreted as the primary strand of the Rose Canyon system at this fault latitude as it is best expressed in the geomorphology. Thus, from the perspective of earthquake recurrence, there appears to have been a cluster of events in the early Holocene followed by many thousands of years of inactivity and soil development until this recent surface rupture.

The timing of this most recent event is well-described by Rockwell and Murbach (1996) and Grant and Rockwell (2002) as after AD 1523 and before the construction of the first mission in San Diego in 1769. The construction of the mission precludes a large earthquake after that time, as a surface rupture in San Diego would assuredly have destroyed the mission, and there is no record of such an event. Thus, the event has been assigned a date of ~AD 1650±120 years.

The other aspect of significance with the soil observations is that the most recent rupture broke only the eastern strands of fault 4 (Figure 6). The reconstruction in Figure 8 suggests that this may be the only event to have ruptured these strands between the deposition of units C1 and C2 and the development of the surface soil, as they are all vertically displaced the same amount, within resolution, and this relationship was seen repeatedly in multiple exposures. The map in Figure 7 shows that these strands have produced 3 m of post channel slip, which implies post unit C1. Taken together, these observations argue that all 3 m of this displacement occurred in the most recent event in ca AD 1650. Considering that the channel is offset at least an additional 5.7 m by fault strands that did not rupture in the most recent event, a simple interpretation is that there have been at least two additional, similar-sized (3 m) events after the end of deposition at the site, or sometime after 8.1 ka. Again, these events must have occurred in the early Holocene to allow sufficient time to develop the surface soil that appears to be offset by only the most recent event.

DISCUSSION AND CONCLUSIONS

Paleoseismic work along the onshore Rose Canyon fault zone in the City of San Diego clearly demonstrates that the fault has sustained recurrent Holocene activity (Lindvall and Rockwell, 1995; Rockwell and Murbach, 1996; Grant and Rockwell, 2002). However, more detailed analysis of existing paleoseismic data suggests that the fault underwent a cluster or burst of activity in the early Holocene, followed by a relatively long period of inactivity during which soil formation occurred across the early Holocene fault strands. Figure 9 summarizes the information on the timing and displacement associated with these Rose Canyon events. In this figure, it is assumed that the moderately developed surface soil represents at least 5,000 years of stability and development, and that its offset is the result of only the most recent event because it is so cleanly displaced. In the most recent event, which is dated in La Jolla and downtown San Diego as being in the past few hundred years but pre-mission period, we attribute the entire 3 m of displacement associated with the faults involved in this rupture. If some of the displacement occurred during earlier events, then the measured slip is less for the most recent event but must be increased for earlier ones. In that the gravel-filled channel is displaced by at least an additional 5.7 m, it seems reasonable that this displacement occurred in an additional two events that were similar in size to the most recent event. Alternatively, there may have been many smaller events that cumulatively produced these observed displacements, but that makes the long hiatus in slip that is evident from the soil development even more problematic. It therefore seems simplest to interpret these data as the result of three relatively large (~3 m) displacements in the period after the end of deposition of unit C1 (post-8.1 ka). This implies that these earthquakes were likely at least M7 in magnitude. If these are average values for displacement in these past events, a magnitude of $M_w7.3$ is calculated using the regressions of Wells and Coppersmith (1994).

The earlier events are derived from the trench data in Rose Creek. Two events (gray stars) are interpreted to have occurred after deposition of the gravel channel of unit C1 because at least 5.7 m of displacement can be attributed to the fault strands that were active in this period, as discussed above, and this is nearly twice that which occurred in the most recent event. Alternatively, the penultimate event could have been substantially larger, although this interpretation seems less likely as the Rose Canyon fault is only about 40 km in length if one considers the major step-overs in San Diego Bay and near Oceanside.

Events 4, 5, and 6 are recorded in the floodplain sediments of Rose Creek. The earliest of these (dotted star in Figures 6 and 9) is well-dated at about 9.3 ka (in unit C4). Event 5 (black stars in figures 6 and 9) appears to have broken all fault strands and occurred soon after 8.1 ka. Event 4 (hollow star) also occurred soon after 8.1 ka if Lindvall and Rockwell's (1995) interpretation is correct that the very weak top-soil horizons between deposition of units C1 through C4 represent only a short amount of time.

Considering that the surface soil represents a long period of stability, it is not possible to simply space the timing of all six events equally for the past 9.3 ka. In fact, if the interpretation is correct that the surface soil represents at least 5 ka of development, then five of these events occurred as a cluster in the period between about 9.3 and 5 ka, with an average interval of recurrence of less than 1 ka. This observation of clustering is similar to many faults Worldwide, including the San Jacinto and San Andreas faults in southern California (Fumal et al., 2002; Rockwell et al., 2006, 2008), the Wasatch, New Madrid and Meers fault in the mid-continent region of the US, the Vilarica fault in Portugal (Rockwell et al., 2009), and the North Anatolian fault in Turkey (Okumura et al., 2009 in review). For the San Jacinto record of 18 surface ruptures at Hog Lake (Rockwell et al., 2008, 2009), the fault has mode-switched from clustered to periodic behavior, but without much predictability as to when such a change in behavior will occur.

For the Rose Canyon fault, its Holocene behavior can be interpreted in several ways, each with its own implications for seismic hazard. The average return period for large surface ruptures during the Holocene is about 1800 years if one simply takes the occurrence of six events (5 intervals) in the past 9.3 ka. However, using the limited timing information at hand suggests that five of these events (4 intervals) occurred in a ~3 ka period, yielding a much shorter return period within the cluster of about 800 years. Further, events 2 through 5 may have occurred in as little a couple thousand years if the surface soil represents as much as 5000 years of development. If the fault principally behaves in a clustered seismicity mode, and if the five early Holocene events represent such a cluster, then one must consider the possibility that the recent earthquake of ca. AD 1650 represents a return to activity and is possibly the first in the next cluster of large earthquakes.

In the first case where earthquakes are assumed to be quasi-periodic in their recurrence, the conditional probability of the occurrence of another M7+ Rose Canyon rupture has a likelihood of less than 1% when the lapse time of only a few hundred years is used. In the second scenario, the probability increases dramatically if we have entered another cluster, because the interval between events within a cluster is much shorter than the long-term average. Additional work on the discrete timing of each Holocene surface rupture is warranted to better constrain the behavior of the Rose Canyon fault, as well as to assign more realistic probabilities for the occurrence of future large earthquakes in America's Finest City.

ACKNOWLEDGEMENTS

I thank the many students and colleagues that have worked with me over the years on the geology of the Rose Canyon fault, including but not limited to (in alphabetical order): Pat Abbott, Mike Hart, Mike Hatch, George Kennedy, Phil Kern, Werner Landry, Tiong Liem, Scott Lindvall, Diane Murbach, Monte Murbach, Scott Rugg, and Dave Schug. Special thanks to Pat Abbott, who tossed me into the Rose Canyon fault activity controversy on my first days in San Diego (little did I know then…).

Table 1. *Descriptions of three pedons of the surface soil capping the Holocene terrace at Rose Creek, as described in the trenches excavated by Lindvall and Rockwell (1995). Refer to Soil Survey Staff (1992) for nomenclature. SDI refers to the soil*

HORIZON	DEPTH (CM)	COLOR (DRY AND MOIST)	TEXT	STRUCTURE	CONSIS
RCF 1					
A1	0–27	10YR4/3; 10YR2.5/2	L	m-1msbk	h; so, po-ps
A2	27–55		L		
E	55–65	10YR4.5/3; 10YR3/3.5	SiL	m-1msbk	h; so, po-ps
Bt	65–94	10YR4/5; 7.5YR3/4	CL	2mabk	vh-eh; s, p
BC	94–115	10YR5/5; 10-7.5YR4/4	SiL	1-2mabk	sh-h; s, p
Cox	115–130	7.5YR5/5, 7.5YR3/4	L	2msbk	h; ss, ps-p
C2	130–154	10-7.5YR5/6; 10-7.5YR3.5/4	SiL	m-1msbk	vh; ss-so, ps
C3	154–180	10YR5/6; 10-7.5YR3/4	L	m-1msbk	vh; so, ps
C32	180–205		L		
C33	205–225+		L		
RCF 2					
A1	0–20	10YR4.5/3; 10YR3/2.5	L	m-1msbk	sh; ss, ps
A2	20–40		SiL		
E	40–43	10YR5/4; 10YR3/3.5	L	m-1msbk	h; so, po-ps
Bt1	43–61	10-7.5YR4/4; 7.5YR3/4	CL	2m-csbk	vh-eh; vs, p
Bt2	61–73	10-7.5YR5/4; 10-7.5YR4/4	L	m-2msbk	vh; s, p
BC	73–83	10-7.5YR5/4; 10-7.5YR4/4	L	m-1msbk	vh; ss, p
C1	83–113	10-7.5YR4.5/4; 10-7.5YR3/4	L	m-1msbk	h; so, po-ps
C2	113–143	10-7.5YR4/5; 10-7.5YR3/4	SiL	m-1msbk	h; so, ps-p
C2	143–173+		L		
RCF 3					
A	0–36	10YR4.5/3; 10YR3/2.5	L	m-1msbk	sh-h; so, ps-po
Bj	36–74	10-7.5YR5/5; 10-7.5YR4/4	Cl	2m-cabk-sbk	vh; s, p
BC	74–105	10-7.5YR 5/6; 10-7.5YR4/4	L	m-1msbk	h-vh; so-ss, ps
C	105–160+	10YR5/6; 10-7.5YR4/4	L	m-1msbk	vh; so, ps

development index of Harden (1982). HI refers to the horizon index values for each horizon. B.D. is bulk density of the soil horizon.

CLAY FILMS	BOUND	% SAND	% SILT	% CLAY	B.D.	H.I.
	g, s	50.28	36.96	12.76	1.49	0.15
		40.01	48.17	11.82	1.32	0.15
	a-c, w	36.65	52.77	10.57	1.31	0.17
2npf; 2npo, v1mkpo	g, s	29.93	37.41	32.66	1.65	0.55
v1npo		36.14	52.27	11.59	1.43	0.35
		43.46	39.71	16.83	1.51	0.31
		44.36	54.86	0.78	1.56	0.28
		38.16	44.9	16.93	1.49	0.26
		38.41	43.77	17.82	1.48	0.26
		28.31	47.95	23.74	1.61	0.26
SDI at 225 cm = 63.2						
SDI at 250 cm = 69.8						
	c, w	37.96	47.12	14.92	1.50	0.16
		39.5	50.53	9.98	1.36	0.16
	a, w	37.36	48.68	13.96	1.57	0.19
2npf, v1mkpf; 2-3npo, v1mkpo	g, s	29.73	37.55	32.72	1.49	0.54
v1-1npf, 1npo	g, s	31.85	43.71	24.44	1.52	0.37
1npf	c, w	38.56	42.53	18.91	1.59	0.33
	g, w	45.17	40.65	14.18	1.43	0.19
		41.97	57.27	0.76	1.35	0.23
		36.82	46.75	16.43	1.30	0.23
SDI at 173 cm = 42.0						
SDI at 250 cm = 59.7						
	a, s	41.81	48.6	9.59	1.10	0.14
1npo	g, s	28.97	40.43	30.6	1.45	0.47
	g, s	46.09	37.71	16.2	1.30	0.26
		39.6	45.18	15.22	1.32	0.26
SDI at 160 cm = 45.7						
SDI at 250 cm = 69.1						

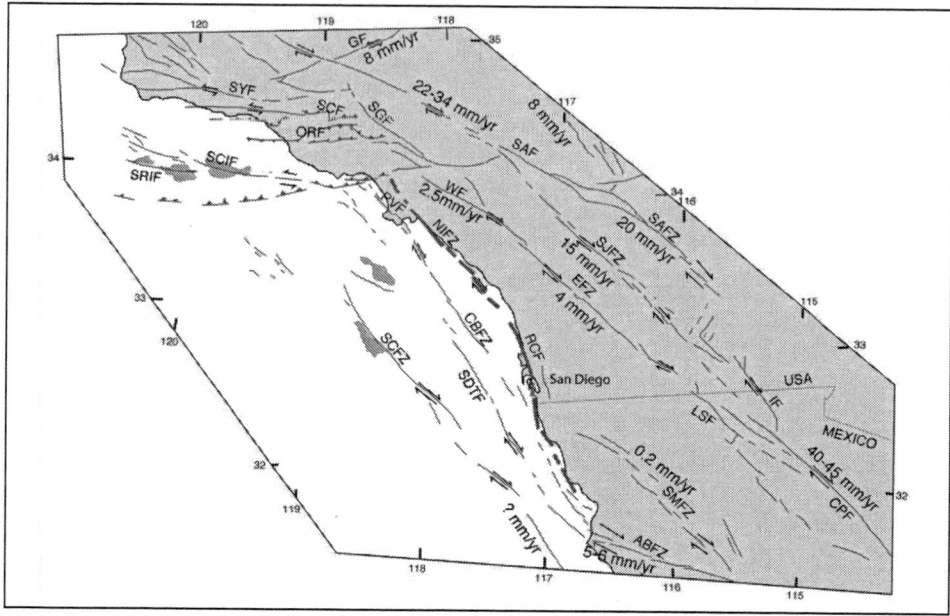

Figure 1. *Generalized map of major faults in southern California, across which there is about 50 mm/yr of relative motion. The Rose Canyon Fault (RCF) is bolded. Other faults are as follows: SAFZ = San Andreas Fault Zone, SJFZ = San Jacinto Fault Zone, EFZ = Elsinore Fault Zone, LSF = Laguna Salada Fault, CPF = Cerro Prieto Fault, ABFZ = Agua Blanca Fault Zone, SMFZ = San Miguel Fault Zone, IF = Imperial Fault, WF = Whittier Fault, PVF = Palos Verdes Fault, ORF = Oak Ridge Fault, SCF = San Cayetano Fault, SGF = San Gabriel Fault, SRIF = Santa Rosa Island Fault, SCIF = Santa Cruz Island Fault, SYF = Santa Ynez Fault, GF = Garlock Fault, SDTF = San Diego Trough Fault, SCFZ = San Clemente Fault Zone, CBFZ = Coronado Banks Fault Zone.*

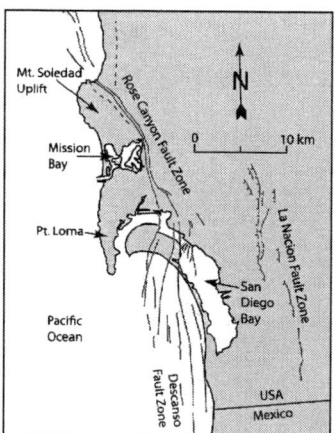

Figure 2. *Elements of the Rose Canyon Fault Zone in the San Diego Region.*

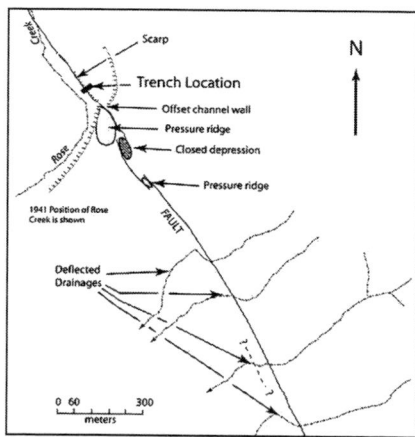

Figure 3. *Tectonic geomorphic features near Rose Creek that are interpreted to be the result of late Quaternary fault activity.*

Figure 4. *Tectonic geomorphology of the Old Town strand of the Rose Canyon fault in Old Town, San Diego. Note that the deflected streams incise into the last interglacial marine terrace (Qt2).*

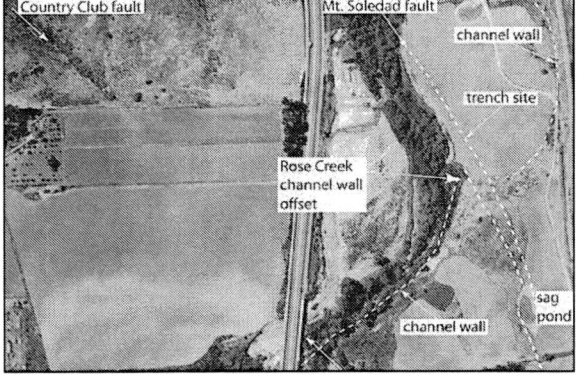

Figure 5. *1928 aerial photograph (Fairchild collection) showing the location of the Lindvall and Rockwell (1995) trench site on the Mt. Soledad strand. Note the vegetation lineament across the Holocene terrace to Rose Creek, suggesting that the Country Club fault may be active in this area. The Rose Canyon fault (sensu stricto) is off the photograph to the northeast.*

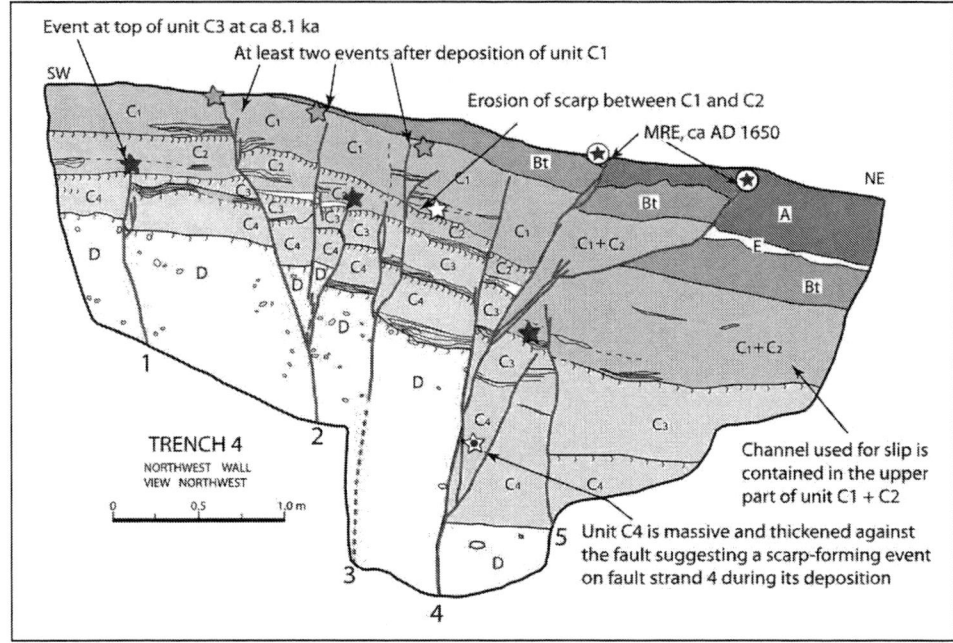

Figure 6. *Log of trench T-4, redrawn from Lindvall and Rockwell (1995). The stars represent evidence for past surface ruptures. The hatcher marks represent weakly formed topsoil (A) horizons. The upper, strongly formed soil is represented by the A, E and Bt horizons. MRE = most recent event.*

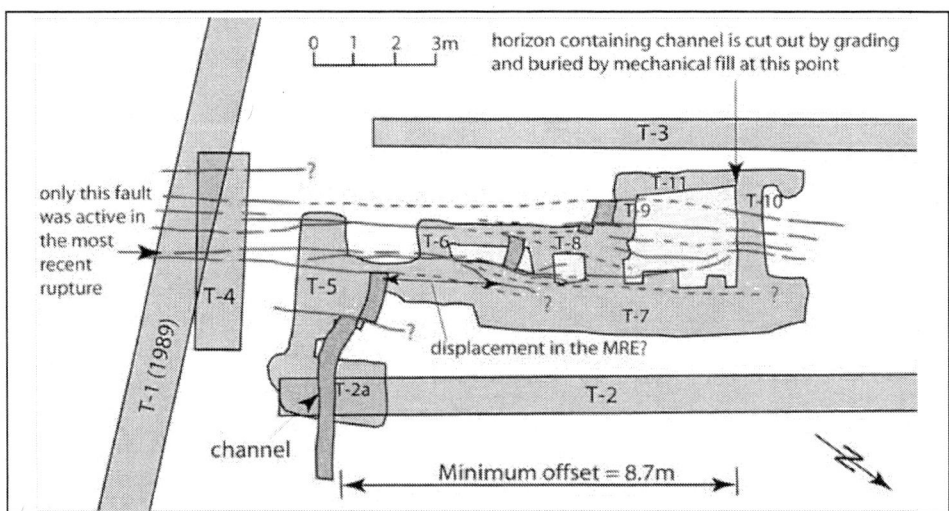

Figure 7. *Map of trenches that were used to map out the extent of the gravel channel embedded in units C1 and C2 (in dark gray). Note that the stratigraphic section that contained the gravel was graded out on the west side of the fault, presumably during grading of the site in 1960. Consequently, the lateral displacement of 8.7 m is a minimum value. MRE = most recent event.*

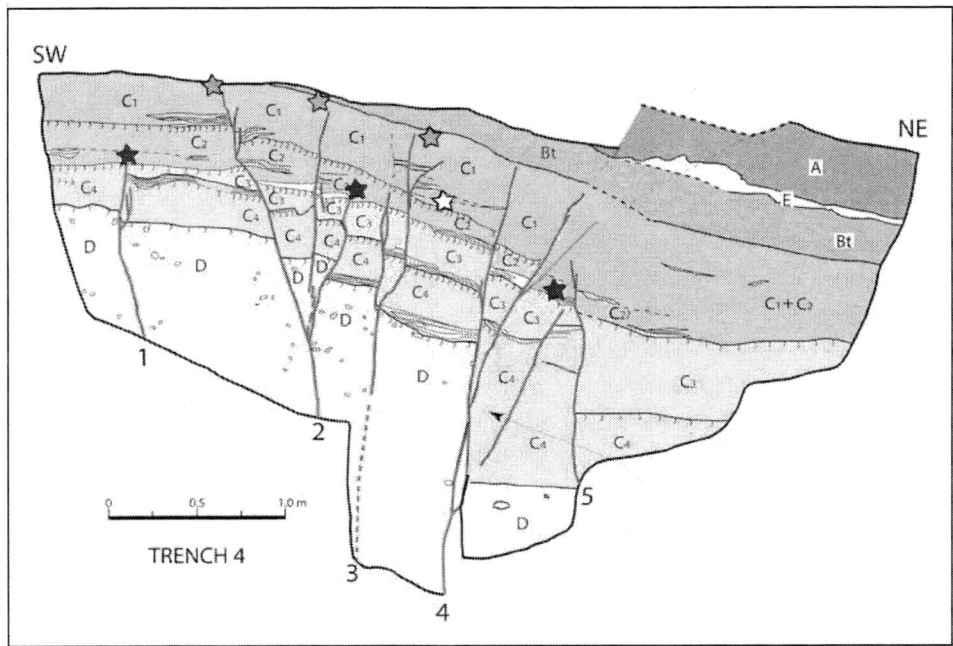

Figure 8. *Reconstruction of vertical separation from the most recent event, as exposed in trench T4. Note that the dominant sense of slip is right-lateral, as exposed in the lateral trench excavations of figure 7, so units may not match perfectly. Nevertheless, units C1 and C2 realign very well when the vertical separation associated with the modern soil is removed, suggesting that this represents a single displacement after the end of unit C1 deposition.*

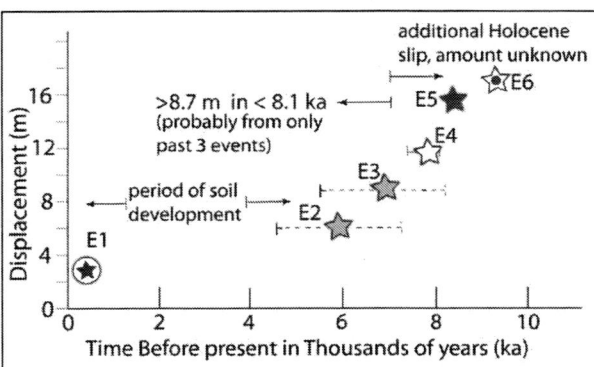

Figure 9. *Time history of surface ruptures interpreted from the paleoseismic information collected for the Rose Canyon fault in the City of San Diego. Six surface ruptures are interpreted from the trench exposures in Rose Creek (using data collected in the 1992 excavations of Lindvall and Rockwell, 1995), with the most recent event date from La Jolla and downtown San Diego (Rockwell and Murbach, 1996; Grant and Rockwell, 2002). The stars correspond to the event horizons identified in figure 6. The displacement values are derived from figure 7, where displacement is only known for the past three events. The strongly-formed soil is interpreted to represent at least 5 ka of development time, although there are no direct measurements to corroborate this estimate. It is possible that additional ruptures have occurred at this site for which there is no evidence due to lack of deposition, so the inferred recurrence interval is a maximum value.*

REFERENCES

Boettcher, R.S., 2001, Foraminifera report, Bayview Reservoir-Kate Sessions Park, La Jolla, California. Unpublished paleontological report prepared for San Diego Paleontological Associates, La Mesa, by Micropaleo Consultants Inc., Encinitas, California, 3 p.

Deméré, Thomas A., 1982, Review of the lithostratigraphy, biostratigraphy and age of the San Diego Formation, in Geologic Studies in San Diego (Patrick L. Abbott, ed.),San Diego Association of Geologists Fieldtrip, April, 1982, p. 127–134.

Ehlig, P., 1980, Rose Canyon Fault. Unpublished report for Southern California Edison, January, 1980, 32 p.

Fumal, T.E., Weldon, R.J., Biasi, G.P., Dawson, T.E., Seitz, G.G., Frost, W.T., and Schwartz, D.P., 2002, Evidence for large earthquakes on the San Andreas fault at the Wrightwood, California, paleoseismic site: AD 500 to present. *Bulletin of the Seismological Society of America*, v. 92, no. 7, p. 2726–2760.

Grant L. B. and Rockwell, T. K., 2002, A northward propagating earthquake sequence in coastal southern California: *Seismological Research Letters*, v. 73, no. 4, p. 461–469.

Gingery, James R., Rugg, Scott H., Hilton, Bruce, and Rockwell, Thomas K., 2010, Fault hazard characterization for a transportation tunnel project in Coronado, California. Fifth International Conference on Recent Advances in Geotechnical Earthquake Engineering and Soil Dynamics, May 24-29, 2010, San Diego, California, Paper No. 7.02C, 13 pp.

Harden, J., 1982, A quantitative index of soil development from field descriptions: Examples from a chronosequence in central California. *Geoderma*, v. 28, p. 1–28.

Kennedy, M.P., 1975, Geology of the western San Diego metropolitan area, California, in Geology of the San Diego Metropolitan Area, California. *California Division of Mines and Geology, Bulletin 200*, p. 9–39.

Kern, J.P. and Rockwell, T.K., 1992, Chronology and deformation of Quaternary marine shorelines, San Diego County, California: in Quaternary Coasts of the United States: Marine and Lacustrine Systems: *Society of Economic Paleontologists and Mineralogists Special Publication No. 48*, p. 377–382.

Kies, R.P., 1982, Paleogeography of the Mt. Soledad Formation west of the Rose Canyon fault, in Geologic Studies in San Diego (Patrick L. Abbott, ed.), San Diego Association of Geologists Fieldtrip, April, 1982, p. 1–11.

Kling, S.S., 2001, Calcareous nannoplankton report, Bayview Reservoir-Kate Sessions Park, La Jolla, California. Unpublished paleontological report prepared for San Diego Paleontological Associates, La Mesa, by Micropaleo Consultants Inc., Encinitas, California, 3 p.

Ku, T.-L. and Kern, J.P., 1974, Uranium-series age of the upper Pleistocene Nestor terrace, San Diego, California, *Geological Society of America Bulletin*, v. 85, p. 1713–1716.

Legg, M.R., 1985, Geologic structure and tectonics of the inner continental borderland off shore northern Baja California, Mexico. Unpublished Ph.D. thesis, University of California, Santa Barbara, 410 p.

Lindvall, S. and Rockwell, T.K., 1995, Holocene activity of the Rose Canyon fault, San Diego, California: *Journal of Geophysical Research*, v. 100, no. B12, p. 24121–24132.

Moore G.W. and Kennedy, M.P., 1975, Quaternary faults at San Diego Bay, California, *U.S. Geological Survey, Journal of Research*, v. 3, no. 5, p. 589–595.

Okumura, K., Rockwell, T.K., Akciz, S., Wechsler, N., Disekai, Aksoy, E., 2009, Slip history of the 1944 rupture segment on the North Anatolia fault near Gerede, Turkey: Constraints on earthquake recurrence models. EOS, December, 2009.

Rockwell, T.K., 2000, Use of soil geomorphology in fault studies: in Quaternary Geochronology: Methods and Applications, J.S.Noller, J.M. Sowers, and W.R. Lettis, eds, *AGU Reference Shelf 4*, American Geophysical Union, Washington D.C., p. 273–292.

Rockwell, T.K., Johnson, D.L., Keller, E.A. and Dembroff, G.R.,1985, A late Pleistocene-Holocene soil chronosequence in the central Ventura Basin, Southern California, U.S.A.: in K. Richards, R. Arnett, and S. Ellis (eds.), *Geomorphology and Soils*, George Allen and Unwin, p. 309–327.

Rockwell, T.K. and Murbach, M., 1996. Holocene earthquake history of the Rose Canyon fault zone, *U.S Geological Survey Final Technical Report* for Grant No. 1434-95-G2613, 37 p. + Appendix.

Rockwell, T., Seitz, G., Dawson, T., and Young, J., 2006, The long record of San Jacinto fault paleoearthquakes at Hog Lake: Implications for regional strain release in the southern San Andreas fault system. *Seismological Research Letters*, v. 77, no. 2, p. 270.

Rockwell, T.K., 2008, Observations of Mode-Switching From Long Paleoseismic Records of Earthquakes on the San Jacinto and San Andreas Faults: Implications for Making Hazard Estimates from Short Paleoseismic Records. *International Geologic Congress*, Oslo, Norway.

Rockwell, T., J. Fonseca, C. Madden, T. Dawson, L. A. Owen, S. Vilanova, and P. Figueiredo, 2009, Paleoseismology of the Vilariça Segment of the Manteigas-Bragança Fault in Northeastern Portugal. Reicherter, K., Michetti, A.M. and Silva, P.G. (eds) *Paleoseismology: Historical and Prehistorical Records of Earthquake Ground Effects For Seismic Hazard Assessment. The Geological Society of London Special Publications*, 316, 237-258. DOI: 10.1144/SP316.15

Shackleton N.J. and Opdyke, N.D., 1973, Oxygen isotope and paleomagnetic stratigraphy of equatorial Pacific core V28238: Oxygen isotope temperatures and ice volumes on a 10^5 and 10^6 year scale, *Quaternary Research*, v. 3, p. 39–55.

Wells, D. L., and K. J. Coppersmith (1994). New empirical relationships among magnitude, rupture length, rupture area, and surface displacement, *Bulletin of the Seismological Society of America*, v. 84, 974–1002.

Adapted from: Fifth International Conference on Recent Advances in Geotechnical Earthquake Engineering and Soil Dynamics and Symposium in Honor of Professor I.M. Idriss, May 24–29, 2010, San Diego, California, Paper No. 7.06c

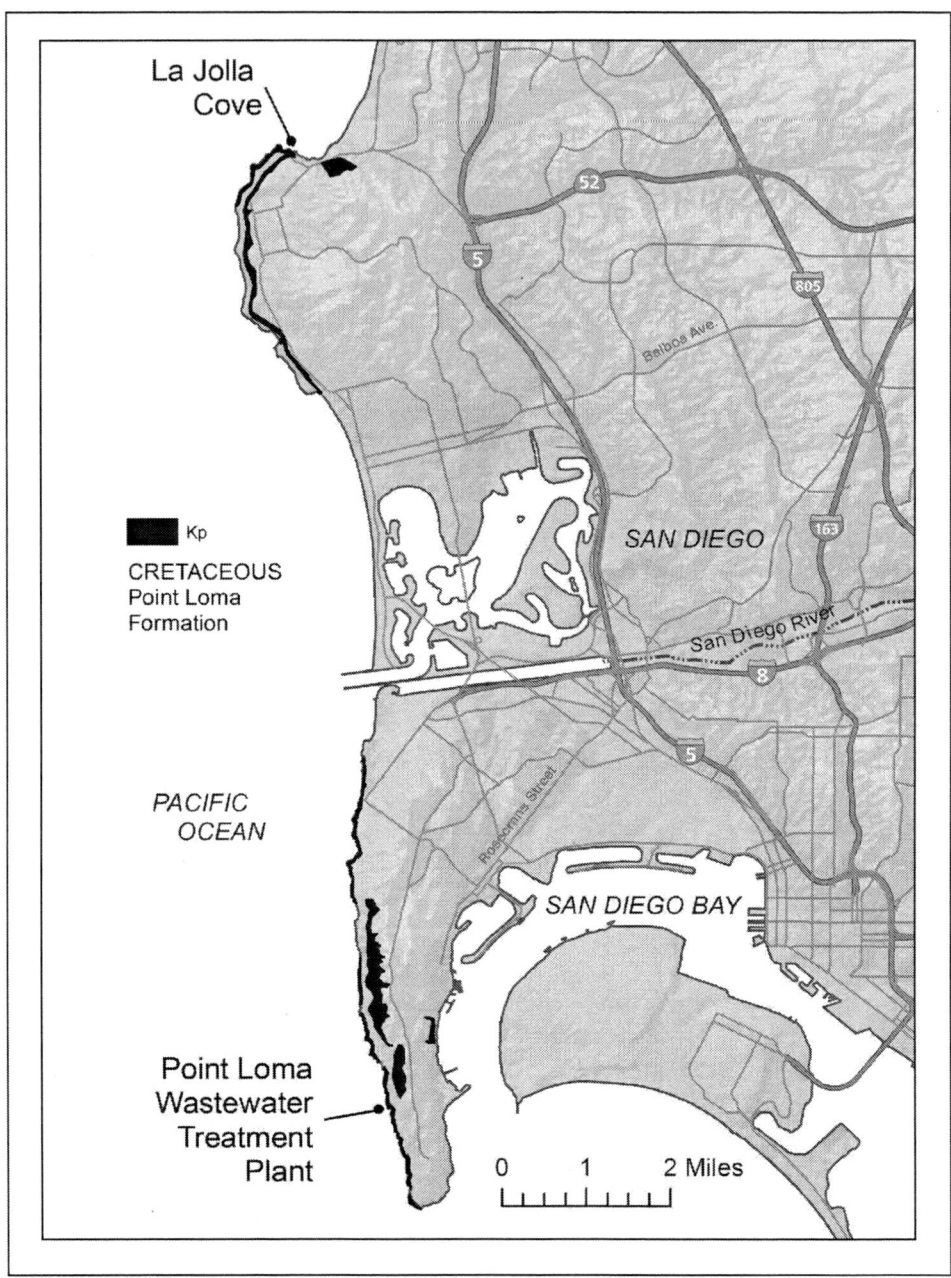

Upper Cretaceous (Campanian-Maastrichtian) Point Loma Formation (Kp), part of the Rosario Group as described by Kennedy (1975) are "interbedded fine grained dusky-yellow sandstone and olive-gray clay shale that occur in graded beds about 30 cm thick." Shoreline caves have formed in the Point Loma Formation as described in Spaulding and Crampton (this volume).

SPELUNKING ON SAN DIEGO'S COASTLINE

Gregory A. Spaulding, P.G., C.E.G., C.H.G.
TerraCosta Consulting Group, Inc.
3890 Murphy Canyon Road, Suite 200, San Diego, CA 92123
gspaulding@terracosta.com

Walter F. Crampton, R.G.E., R.C.E.
TerraCosta Consulting Group, Inc.
3890 Murphy Canyon Road, Suite 200, San Diego, CA 92123
wcrampton@terracosta.com

DISCLAIMER

Numerous sea caves punctuate the California coastline, contributing to the many geologic hazards that should be respected when conducting investigations. All of the sea cave investigations described in this paper were undeniably hazardous, with three of the four sea caves accessible only by boat. The authors investigated these sea caves because they were paid to do so and knowingly accepted the hazards as part of the job. Access to one of the sea caves resulted in several broken ribs and numerous lacerations. Again, these injuries were accepted as part of the job. We strongly recommend that any recreational access to these sea caves be avoided and as with any cave diving, any attempt at access should only be made by experienced spelunkers and include contingency plans, and standby safety personnel on a boat with a marine radio able to contact the proper authorities, should the worst happen. Entrance to several of these caves was no sh_t dangerous and should not be attempted unless there is a damn good reason to do so!

INTRODUCTION

San Diego's coastline consists of a combination of sea cliffs and lowlands, the latter occurring where valleys enter the sea and estuaries are cut off by barriers (Kuhn and Shepard, 1984). The cliffed sections are generally formed where the more erosion-resistant Tertiary and Cretaceous bedrock has been uplifted and exposed to the sea, with the highest portion in the Torrey Pines area, where some sections reach well over 300 feet in height. While in general, sea cliff erosion along San Diego's coastline is a result of wave attack at the base of a coastal bluff, other geologic factors come into play, which affect the morphology and rate of erosion locally. Lithology, bedding, groundwater seepage, and more commonly jointing and faulting can all play a significant role in the various features formed by erosion within the bluffs.

A common characteristic of the sea cliffs at Point La Jolla and Point Loma, and, to a lesser extent, the Eocene-age sea cliffs in North County San Diego, is the formation of sea caves. The sea caves, many of which extend inland many tens to hundreds of feet from the cliff face, are primarily formed by wave action. The wave energy is concentrated and funneled along prominent joint sets or faults in the cliff rock, gradually widening these joints to form surge channels and sea caves (Kennedy, 1973). Over time, wave-induced erosion progresses headward, laterally, and upward in the formation of sea caves, leading to uniform hollowing and eventual roof collapse. As marine erosion continues, the entire seaward face of the sea cliff erodes away, eventually leaving the more resistant undisturbed rock some distance from the joint or fault lineament to form the sidewalls of a small lineament-controlled cove (Photo 1). This paper discusses four notable sea caves that have formed in San Diego's Cretaceous-age Point Loma Formation.

GEOLOGIC SETTING

Both the Point La Jolla and Point Loma coastal bluffs are underlain by marine, upper Cretaceous, and Pleistocene sedimentary rocks. The upper Cretaceous rocks of the Point Loma Formation crop out continuously along the lower portion of the westerly facing coastline, generally forming the lower cliffed section of the bluffs. The upper flat-lying Pleistocene deposits belonging to the Bay Point Formation have been deposited on the now-elevated 80,000 to 120,000-year-old abrasion platform and consist of a poorly consolidated sandstone. The rocks at Point La Jolla and Point Loma are lithologically the same and lie at a similar stratigraphic position (Kennedy, 1973).

All of the sea caves observed in our study areas formed within the upper Cretaceous Point Loma Formation. Caves were mainly observed forming along en echelon faults or joint sets that likely formed within the Point Loma Formation during the development of the Rose Canyon fault system. While most of these strain features appear to be antithetic to the

Rose Canyon system, some have also formed by strain release due to erosional unloading parallel to the coastal bluffs.

Kennedy (1973) describes the beginning stage of sea cave development as the formation of surge channels. These channels are numerous within the study areas, and channels that follow intersecting joint and fault planes gradually evolve into caves by progressive basal undercutting. Cave excavation and formation are aided by the concentration and agitation of abrasive sand and rock particles along these joints. Eventually, cave erosion progresses until either all or a part of the roof collapses, forming a blowhole and/or natural bridge in the Point Loma Formation.

Eventually, lateral erosion and removal of the bridge leaves the formation of a small cove and pocket beach.

SEA CAVES IN THE LA JOLLA AREA

The City of San Diego has inventoried their coastal bluffs on a regular basis going back to the 1970s. For the past 25+ years, the authors have been intimately involved with inventorying the bluffs, measuring erosion and determining the impact, if any, that the erosion is having on public improvements. In 2002, the City became concerned that some of the sea caves that were developing might collapse or somehow impact shoreline improvements or endanger the public, and commissioned the authors to perform a survey of sea caves between Gold Fish Point and the Children's Pool in La Jolla. The following discussion illustrates one of the more spectacular caves explored and mapped during the study.

The La Jolla coastal bluffs are bordered by a narrow wave-cut Quaternary-age terrace or bench, with elevations ranging from 30 to 80 feet MSL along the top of the bluffs. Wave impact erosion has etched out the less resistant rock along faults and fractures in the coastal bluff resulting in the shallow coves and sea caves, which punctuate the La Jolla coastline. The more resistant rocks of the Point Loma Formation form the lower cliffed section of the coastal bluff and shore platform, which extends seaward. The relatively flat surface of the modern-day abrasion platform is interrupted by isolated erosion-resistant rock, which forms sea stacks and topographic highs. Further seaward, the abrasion platform becomes progressively deeper, and is locally incised by surge channels that have formed along the trends of major joint sets or faults, which affect the erosion resistance of the rock.

One of the City study areas, Area 61B (Figure 1, Photo 2), is located southwesterly of Goldfish Point, along Coast Boulevard from Coast Walk near The Shell Shop northwesterly a distance of approximately 500 feet. Elevations along the top of the bluff range from approximately 45 to 80 feet, MSL. Northeast-trending faults and joint systems have locally weakened the bedrock in the area (Photo 2). Marine erosion (and, to a more limited extent, subaerial erosion) acting along these planes of weakness have formed a rather spectacular cave system within the cove area (Photo 3). A relatively substantial sea cave, formed along a fault (or joint set) known as Cooks Crack, is located near the most northwesterly part of this area. Figures 1 through 5, and Photos 4 through 9, illustrate the extent of the Cooks Crack sea cave.

Numerous sea caves within the Point Loma Formation have formed along faults and joints within the 80 million-year-old geologic unit, many of which extend beneath Coast Boulevard and/or other bluff-top improvements. The formation and growth of sea caves within this geologic unit is quite prevalent in both La Jolla and Point Loma, and while the faults and joints may literally extend for miles, the growth of the sea cave is rather gradual and limited to only the seaward portion of the sea cave, where the hydrodynamic wave forces wedge and cleave off weaker sections of the rock along the fault lineament. As the sea cave grows, surging water can then move chunks of rock, further abrading the base of the cave, developing more linear surge channels in the absence of significant debris and wider sea caves in the presence of more debris.

As these caves form and grow, the stability of the roof rock remains quite high, until the thickness of the roof rock thins and/or the width of the arch becomes large enough to induce tensile failures in the apex of the roof associated with bending stresses induced by the overlying terrace deposits. City Geologists considered there to be a potential risk to Coast Boulevard and suggested that the cave be filled with concrete or a concrete slurry. As we understand, overriding environmental considerations, along with community concern, resulted in deferring any immediate action until additional investigative work could be performed. The authors' contract included the inspection, survey, and systematic evaluation of the Cooks Crack sea cave, with the inspection performed during the tidal low on November 19, 2002.

To assist in the mapping effort, two survey pins were set near the cave entrance for reference. A plumb line was set from these pins for referencing a distance and bearing to map the limits of the cave. Based on the results of our mapping effort, it is apparent that the cave developed along a fault system within the Point Loma Formation that extends under Coast Boulevard. The cave was measured to be over 150-feet long, over 50-feet wide at its maximum width, and over 20-feet high at its maximum height (Figures 1 through 5). From a review of our survey data and information obtained from other sources, we estimate that approximately 15 feet of Point Loma Formation roof rock still remains between the top of the cave and the bottom of the Bay Point Formation. Observations made during our mapping effort revealed a substantial portion of the roof of the cave collapsed at some time in the recent past (Figures 2 through 4; Photos 6 and 7).

In discussions with the City Geologists who conducted a limited survey in 1996, it was concluded that the cave had not substantially changed in the seven years between the two surveys. The 1996 cave survey encountered the same significant debris pile requiring some climbing to scale. Thus, the debris from the roof collapse we observed also existed in 1996. The cave height is consistent and the primary cave chamber is still approximately 15 feet in width. The central lower sidewall openings likely also existed in 1996. The eastern opening is hard to reach and not at all obvious. The western opening, although more accessible, is also not obvious, only 3 feet in height and may have been at least partially, if not totally, buried with sand.

The debris pile, which consists of hard and very large blocks of intact Point Loma Formation, fairly effectively protects the interior portion of the sea cave (Figure 5), and likely contributes to the apparent lack of cave growth. It is possible, however, that the interior of the cave is enlarging, and it should be periodically resurveyed to confirm its apparent stability. Sea cave roof rock stability was only analyzed two-dimensionally, recognizing that, in most cases, the three-dimensional geometry likely further stabilizes the cave roof. The relatively large, intact blockfall within the inner chamber of Cooks Crack was back-calculated assuming a factor of safety of 1.0 and solving for the maximum tensile stress, σ_{xm}, resulting from the pull-apart collapse. Since the average tensile strength is about one-half of the effective stress cohesion intercept (Sitar, et al., 1981), the stability of any roof rock configuration can then be analyzed for a given soil strength.

Stability analyses for the five failure geometries considered in Cooks Crack are summarized on Figure 6.

Given this background, and although stable today, it is still reasonable to conclude that, at some time in the future, the Cooks Crack sea cave will collapse. It is also important to point out that the roof rock stability calculations are based on static conditions and no additional cave growth. Seismic loading increases both the dead weight of the overhanging block and, depending upon the direction of bedrock acceleration, can also significantly increase the pull apart moment, thereby further reducing the stability of the roof rock.

SEA CAVES OF POINT LOMA

In 2002, The City of San Diego and the National Park Service commissioned the authors to perform an evaluation of bluff retreat and encroachment of sea coves and sea caves upon the Point Loma Wastewater Treatment Plant South Access Road (Gatchell Road) and other existing bluff-top improvements (Figure 7, Photo 10).

Our field investigation also included sending swimmers into all of the caves along the study area from a boat. To aid with the mapping, a surveyor was dropped off on a rock (remnant sea stack) to help provide survey control for the swimmers/surveyors. From the floor of these caves, and in particular the north and south coves, the base of the sea cliff was surveyed at the back of the cave/cove, the limits of the sea cave/cove, and the location of any ledge rock encountered above sea level. The results are shown on Figures 8 through 11. The information on these figures provided a baseline for comparison to later surveys, although the X-Cave had been previously surveyed by the authors in 1992.

The complex sea cave (the X-Cave) just north of the south cove (Figures 8 and 10, Photos 11 and 12) appears to be rapidly enlarging since last surveyed in 1992, and will likely reach the landward limit of the underlying sea cave in 20 to 40+ years. At that time, the back and sides of the narrow future cove would be near vertical. The back of the future cove would then retreat at about the same rate as the sea cove to the south.

The sea caves along the center bluff section of the study area (the tunnel cave and the cave to the south) are controlled by bluff-parallel joints, which form an avenue of weakness for marine erosion at the base of the sea cliff.

Section D (Figure 11) shows the roof consisting of substantial thicknesses of erosion-resistant Point Loma Formation. Photos 13, 14, and 15 depict these two caves. Due in large part to the thickness of the roof rock above the tunnel cave, the stability of the cave is quite high.

Rum Runners Sea Cave

The bluff-top public parking lot adjacent to Froude Street overlies a very extensive sea cave formed along two joint sets roughly trending northwest/southeast and northeast/southwest (Figure 12). This cave was reportedly used during the Prohibition by whiskey runners during high tide when a boat could drive into the northeasterly cave entrance (prior to placement of the rock revetment) and tie up to a concrete dock, the remains of which exist today. Concrete steps lead up from the dock to a tunnel that at one time traversed under Sunset Cliffs Boulevard and eventually surfaced on the private property to the east. The tunnel itself was excavated immediately on top of the Point Loma contact, with the entirety of the tunnel being in the Bay Point Formation. Currently, the seaward 20± feet of the tunnel is still accessible; however, the remainder of the tunnel has been filled with concrete (Photos 16 and 17).

The main interior of the cave (that was originally accessed by water) has a relatively tall (20± feet) domed roof, with active roof rock blockfalls still occurring. Eventually, the roof of this cave will collapse, resulting in an appearance similar to the sea cave just north of Monaco Street. Although the authors have inspected the interior of this rather impressive cave on several occasions, a reasonable estimate of the areal extent of the cave is difficult and would be necessary to make a reasonable evaluation of the susceptibility of its collapse and threat to Sunset Cliffs Boulevard.

The N40W fault controlling the subsidiary sea cave off the Rum Runner's Cave also appears to have active roof rock blockfalls still occurring, with subaerial erosion encroaching on the northerly edge of the parking lot where the northerly parking lot guard rail has been moved 10 feet southerly to accommodate the slow, yet persistent, subaerial erosion. Most notable is the significant enlargement of what was at one time a small fissure along the small peninsula that extended to the northwest beyond this cove, where now there is a fairly large arch that has formed with what today is probably 20± feet of roof rock still supporting an arch that is locally 20+ feet wide.

CONCLUSIONS

Literally thousands of sea caves punctuate the Southern California coastline, confined to the lower Eocene and older cliff-forming geologic units. While all sea caves enlarge and will eventually collapse, the vast majority of these sea caves are, at any given time, reasonably stable, given sufficient roof rock and cave geometry to ensure their stability. While the stability of sea caves is important when evaluating the stability of nearby bluff-top improvements, environmental considerations typically prohibit the infilling of sea caves until absolutely necessary. Moreover, at least within the city of San Diego, numerous sea caves do exist under bluff-top structures, where the mouths of these sea caves have been filled to stop ongoing marine erosion, recognizing that the cave itself poses no threat to the stability of these overlying structures. As a result, the public continues to enjoy the benefits of a very dramatic coastal experience.

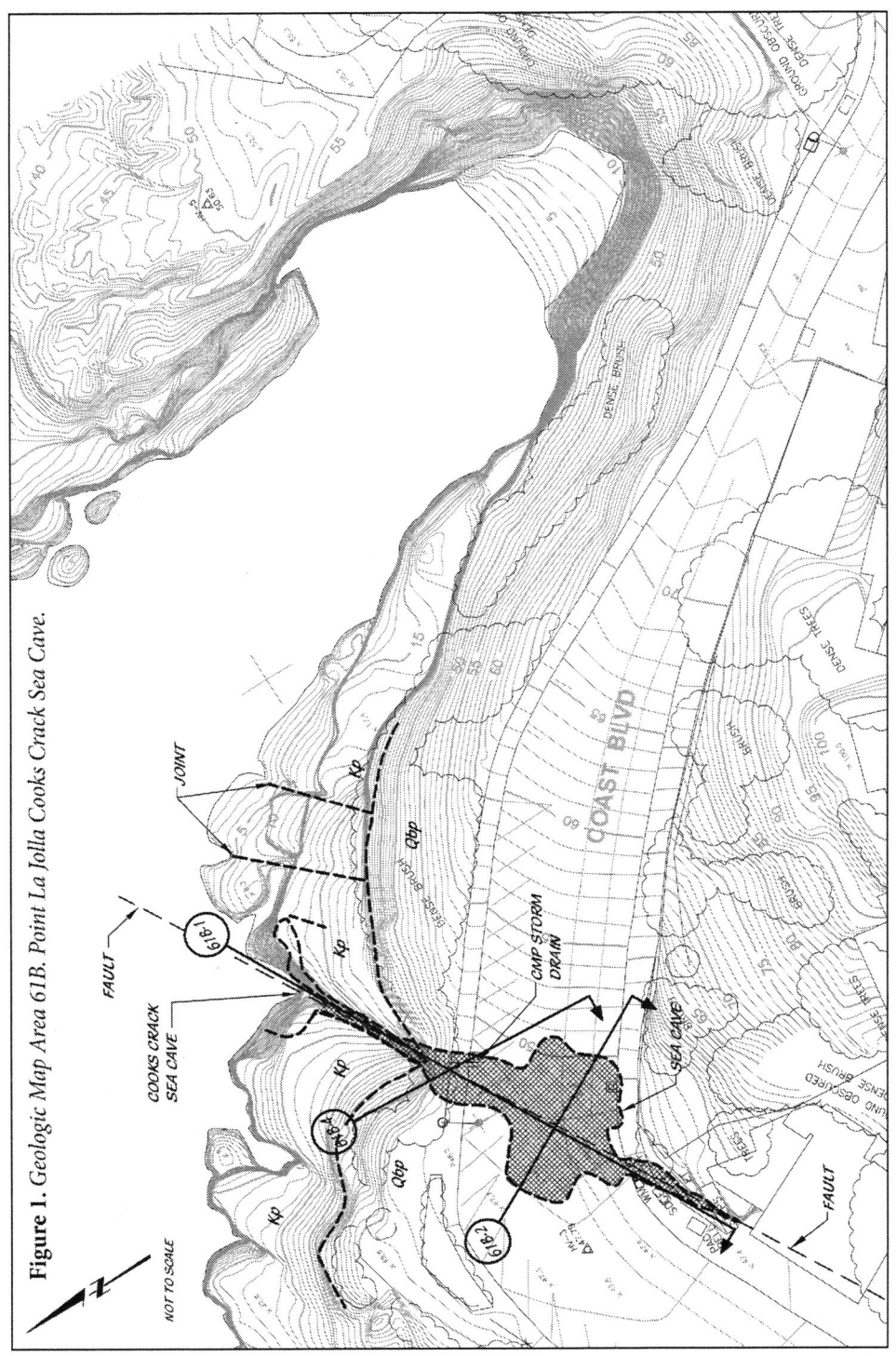

Figure 1. Geologic Map Area 61B. Point La Jolla Cooks Crack Sea Cave.

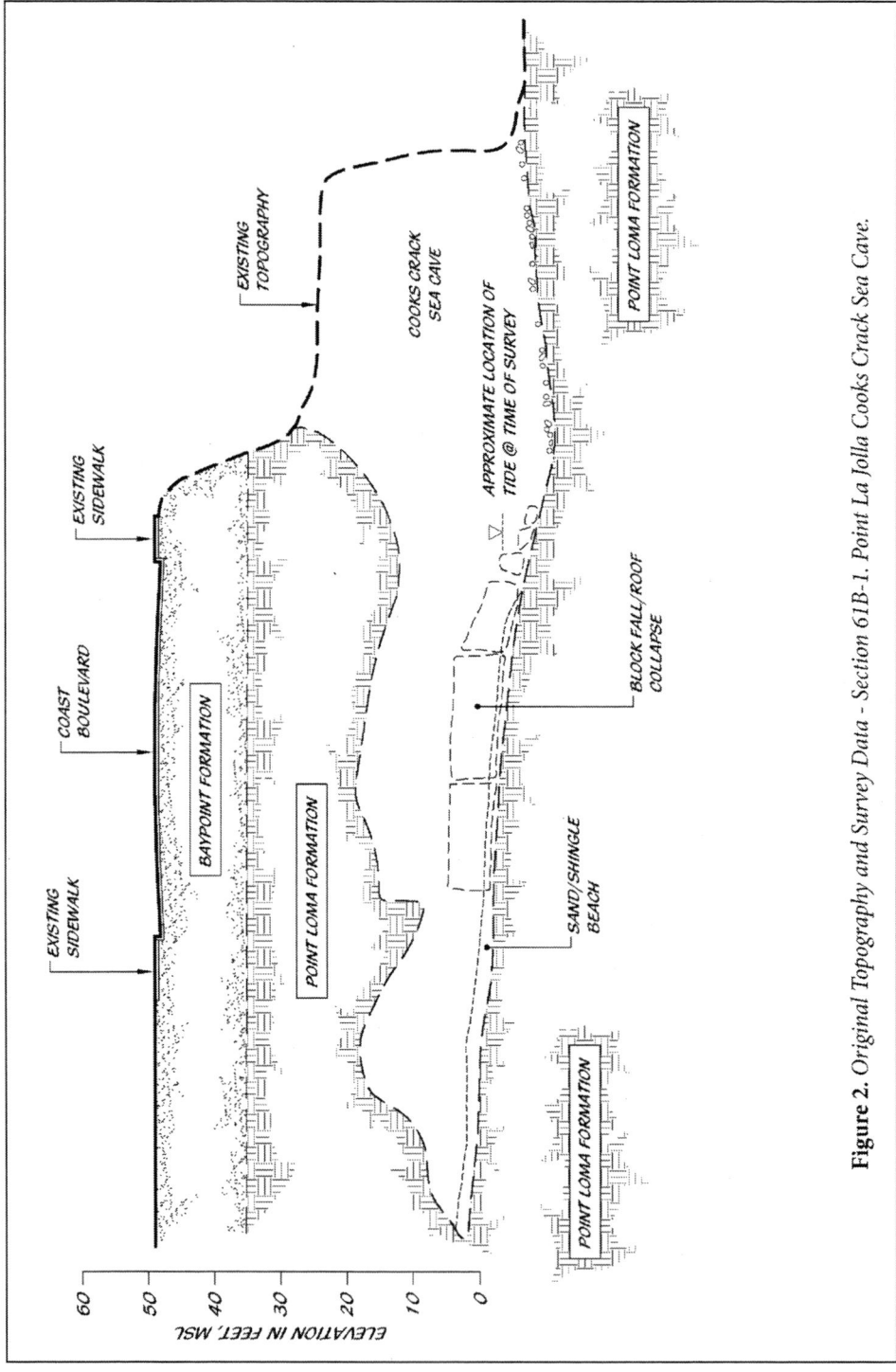

Figure 2. *Original Topography and Survey Data - Section 61B-1. Point La Jolla Cooks Crack Sea Cave.*

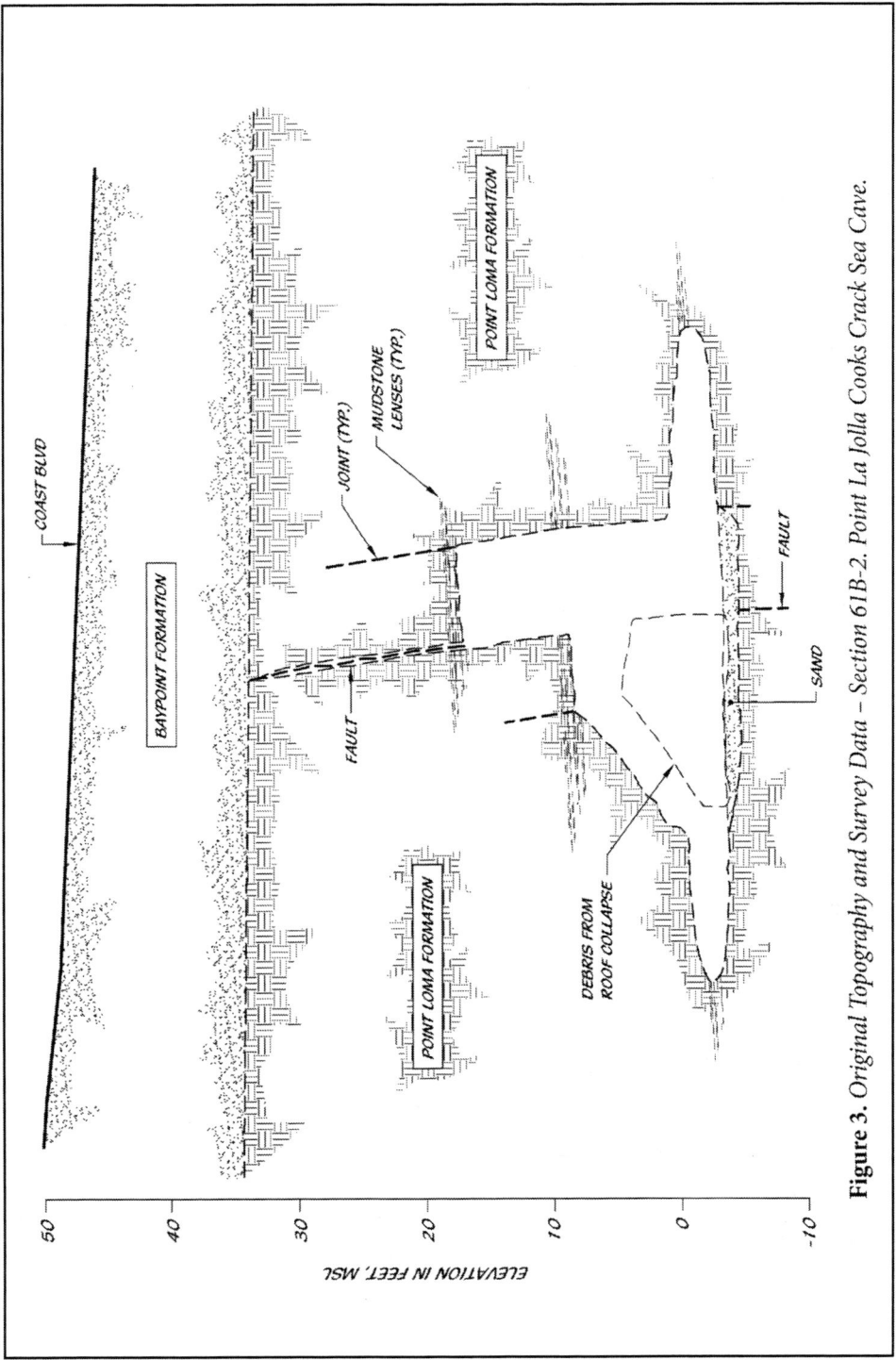

Figure 3. *Original Topography and Survey Data – Section 61B-2. Point La Jolla Cooks Crack Sea Cave.*

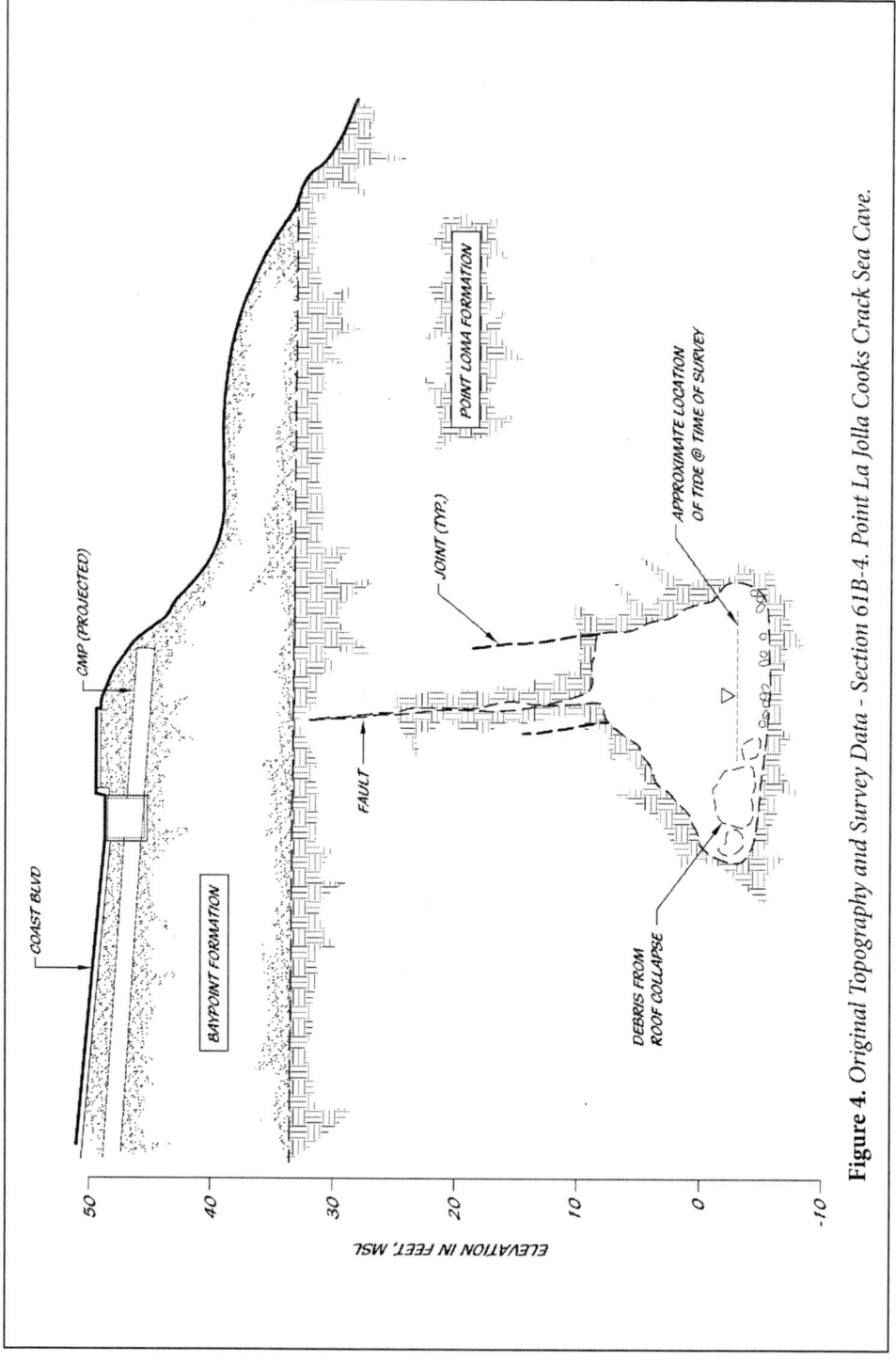

Figure 4. Original Topography and Survey Data - Section 61B-4. Point La Jolla Cooks Crack Sea Cave.

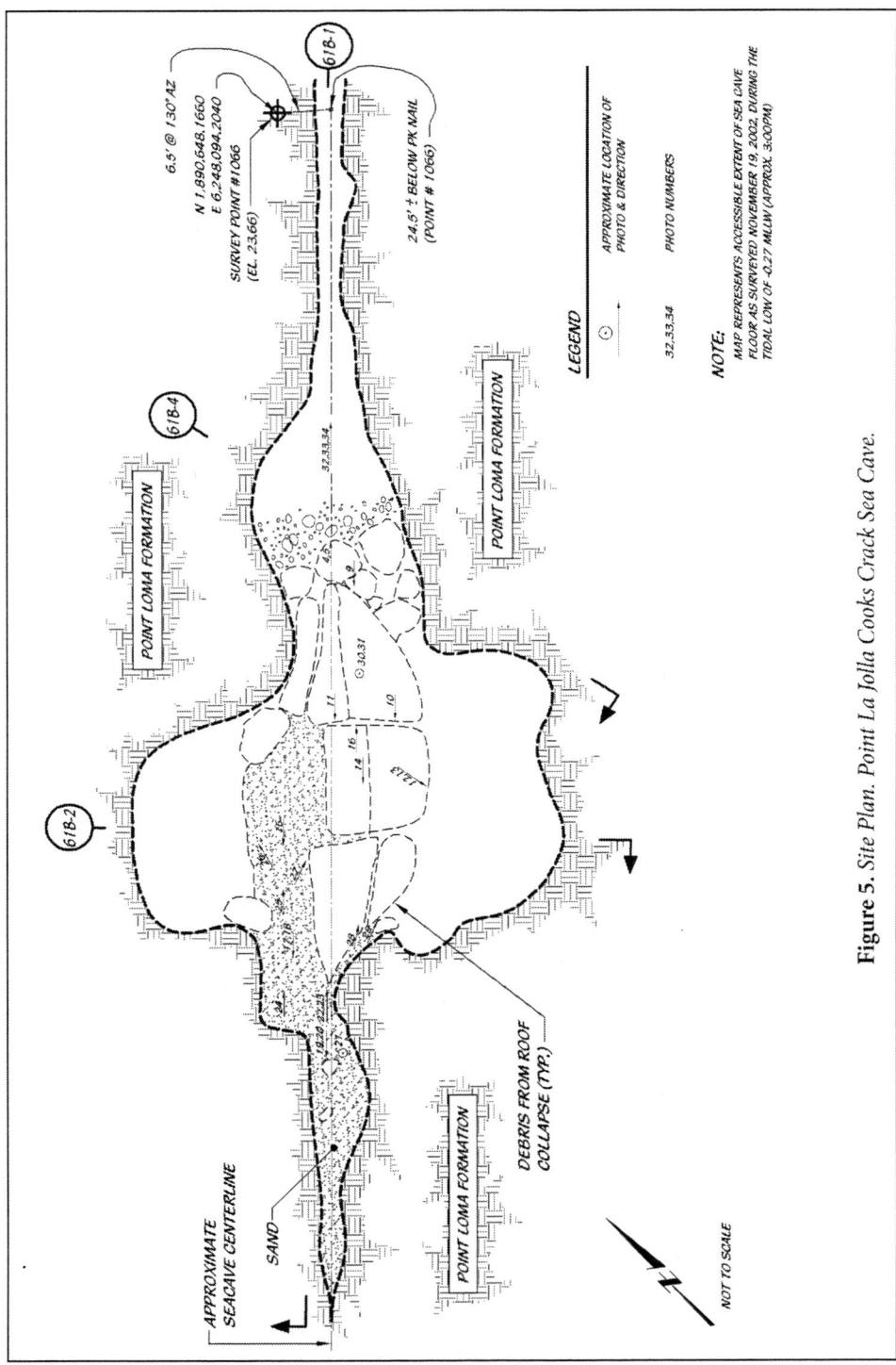

Figure 5. Site Plan. Point La Jolla Cooks Crack Sea Cave.

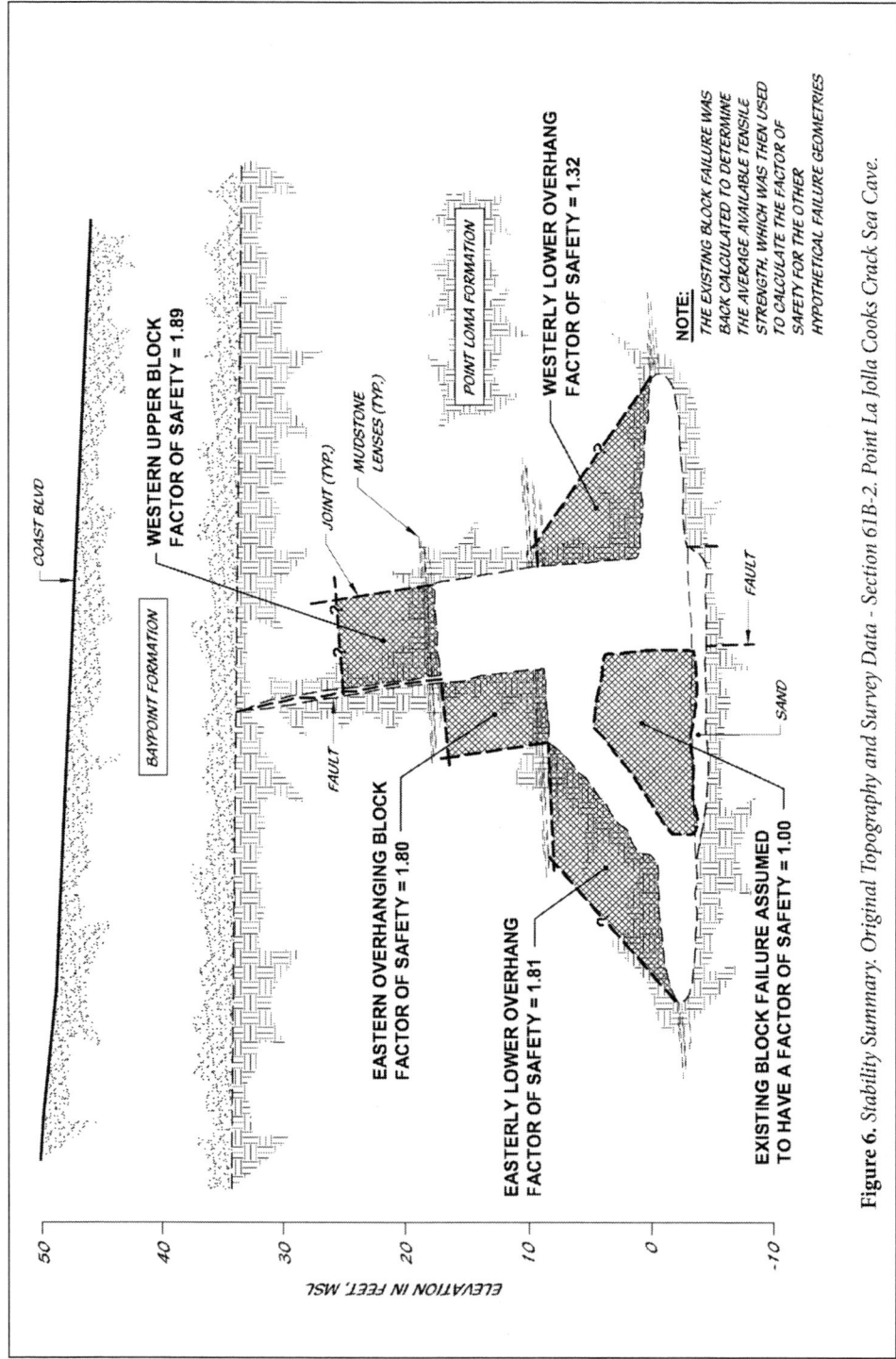

Figure 6. *Stability Summary, Original Topography and Survey Data - Section 61B-2. Point La Jolla Cooks Crack Sea Cave.*

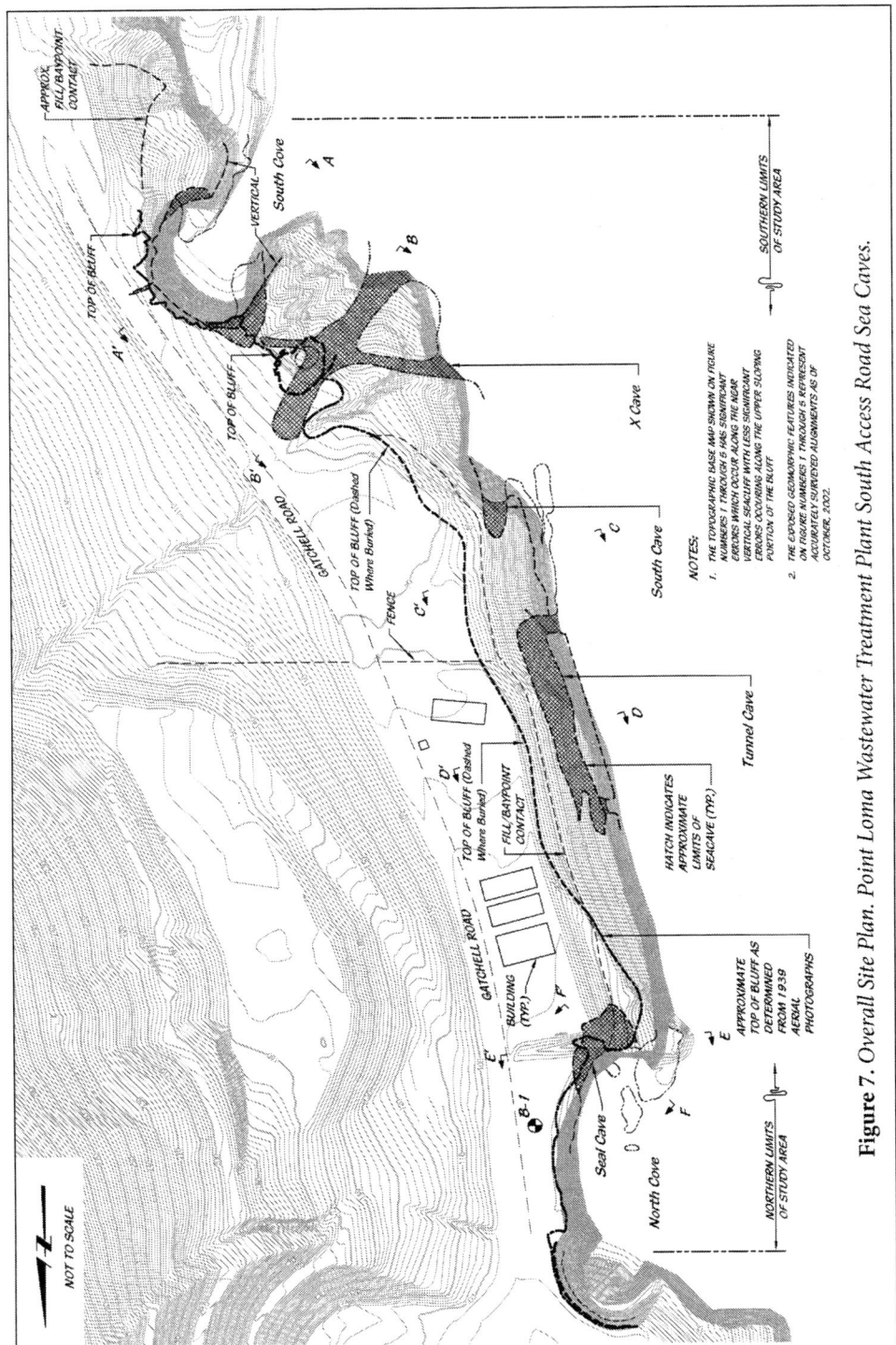

Figure 7. *Overall Site Plan, Point Loma Wastewater Treatment Plant South Access Road Sea Caves.*

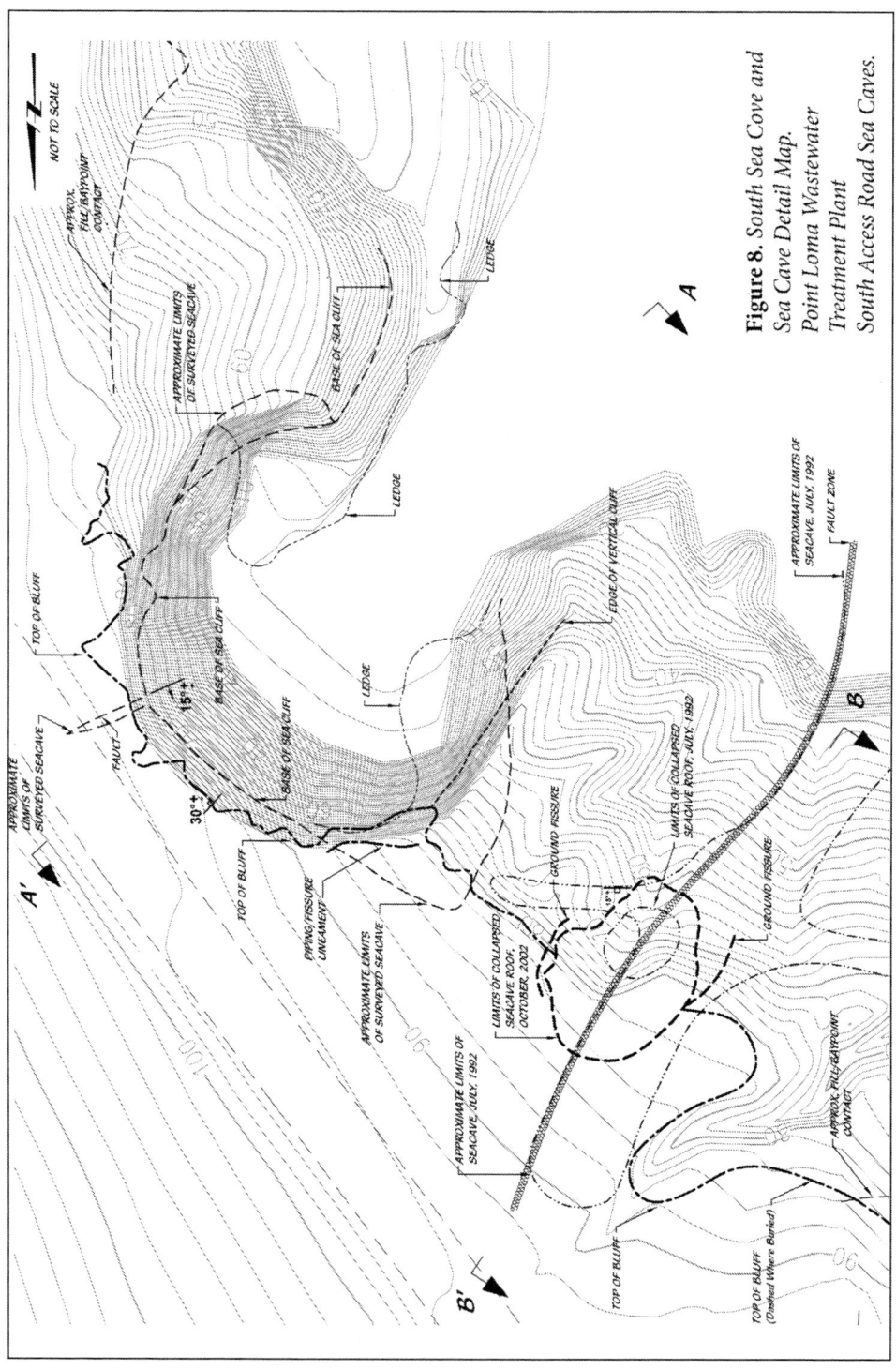

Figure 8. *South Sea Cove and Sea Cave Detail Map. Point Loma Wastewater Treatment Plant South Access Road Sea Caves.*

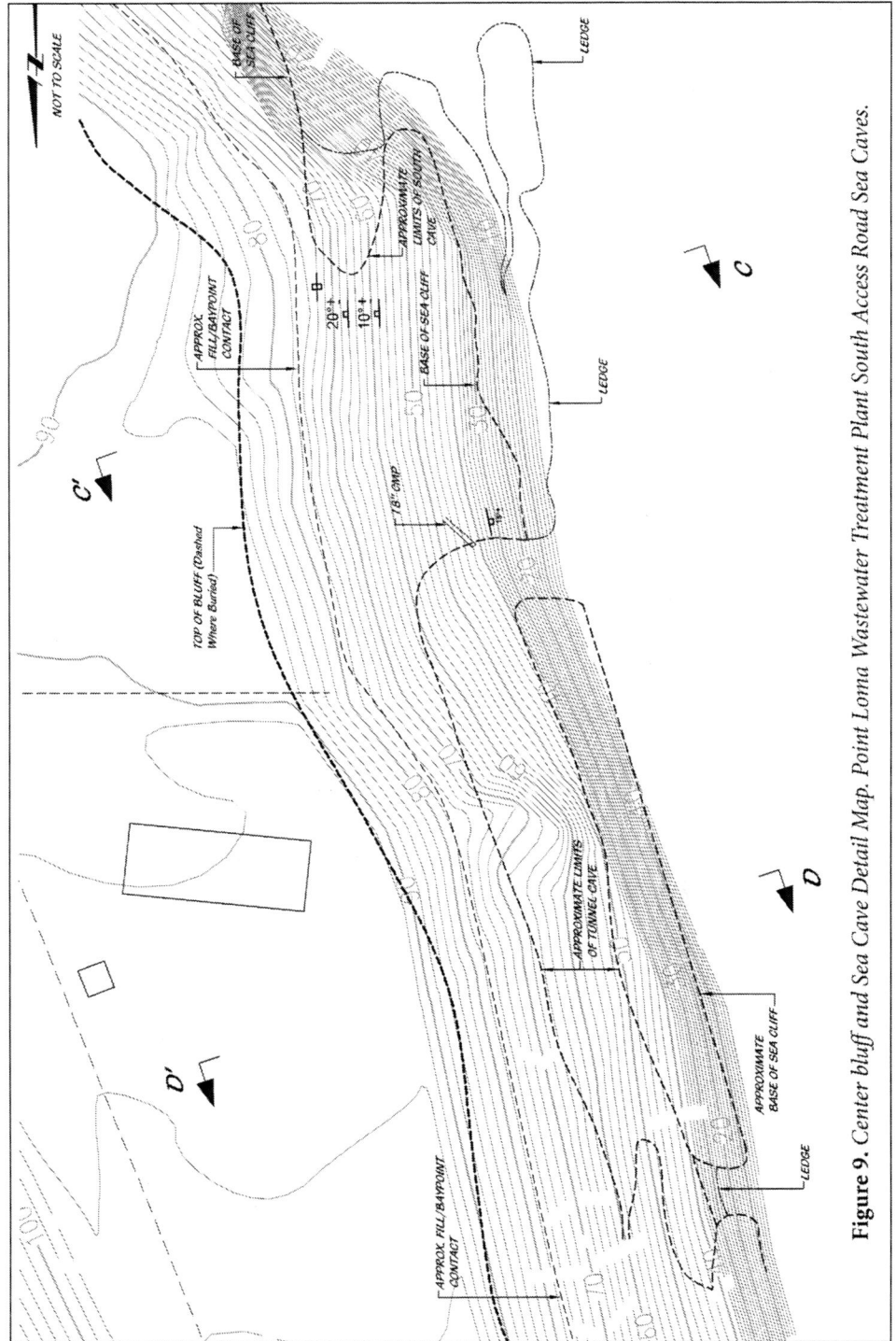

Figure 9. Center bluff and Sea Cave Detail Map. Point Loma Wastewater Treatment Plant South Access Road Sea Caves.

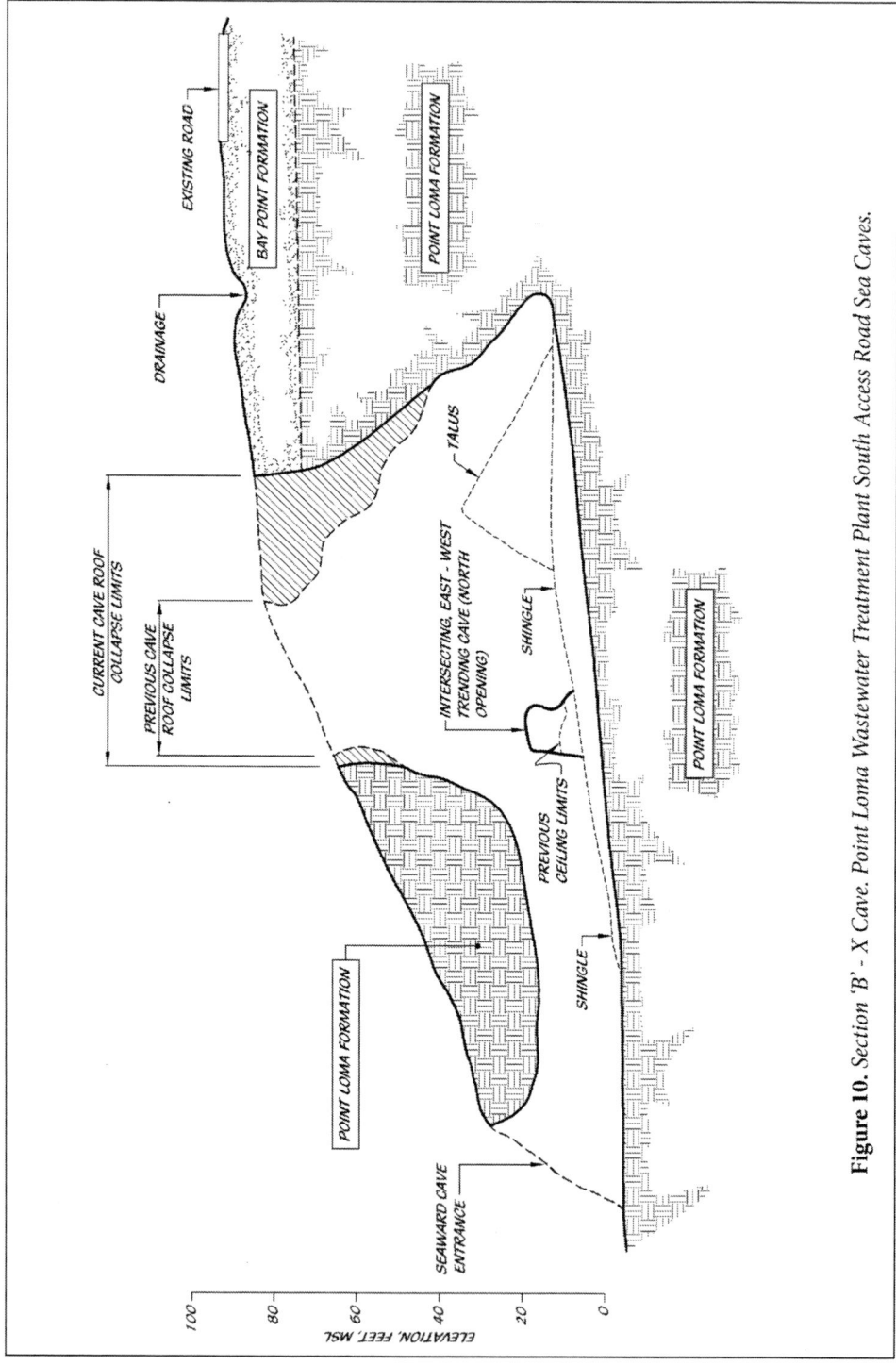

Figure 10. Section 'B' - X Cave. Point Loma Wastewater Treatment Plant South Access Road Sea Caves.

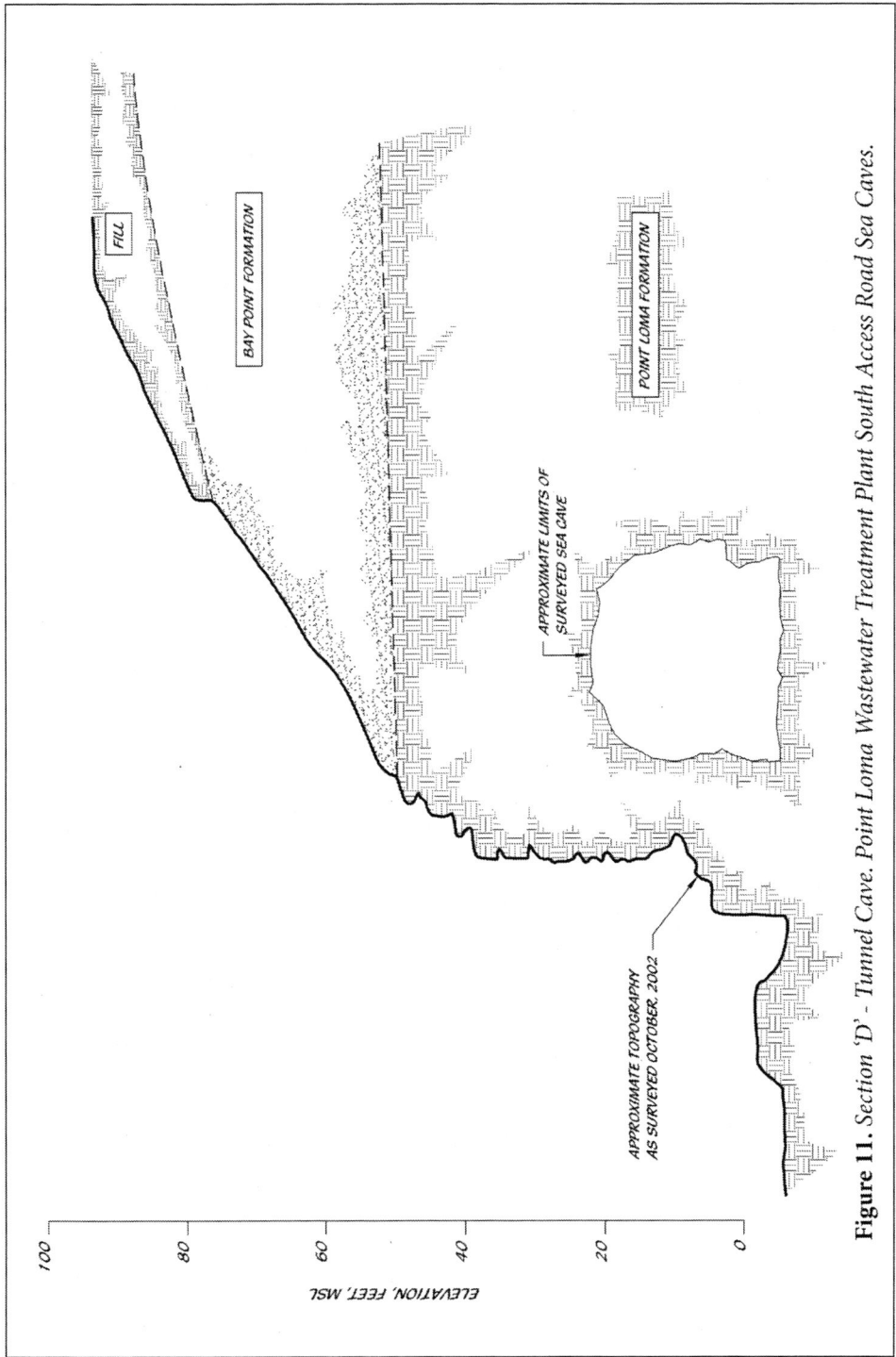

Figure 11. *Section 'D' - Tunnel Cave. Point Loma Wastewater Treatment Plant South Access Road Sea Caves.*

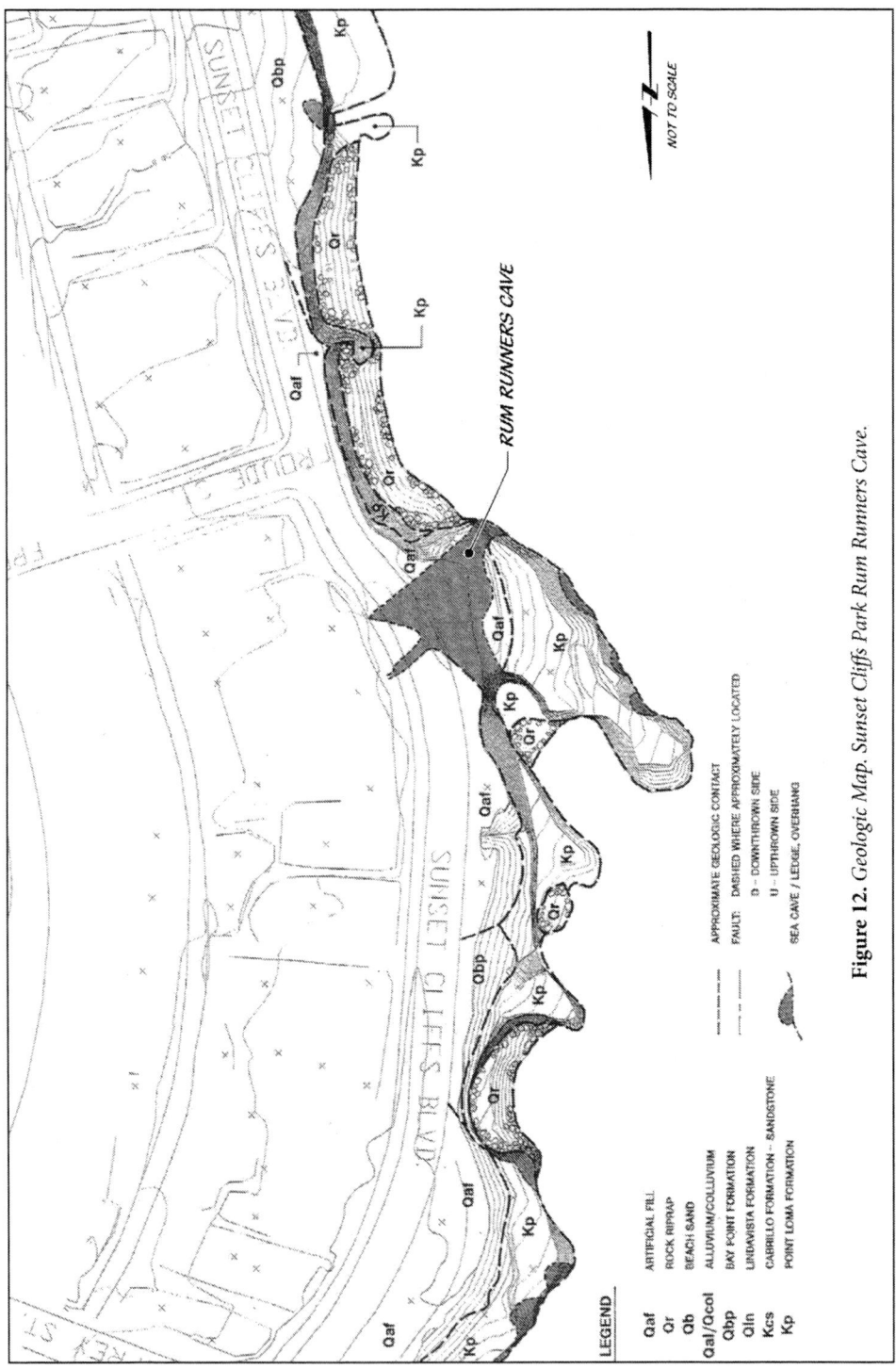

Figure 12. *Geologic Map. Sunset Cliffs Park Rum Runners Cave.*

Photo 1. *Pump Station 35 at Sunset Cliffs Boulevard and Monaco Street.*

Photo 2. *Cooks Crack Sea Cave extends under Coast Boulevard near Goldfish Point in La Jolla, California.*

Photo 3. *Note incipient cave forming along joints and faults.*

Photo 4. *Cooks Crack looking southerly. Note localized erosion in upper bluff.*

Photo 5. *Cooks Crack looking northerly. Note erosion caused by wave spray.*

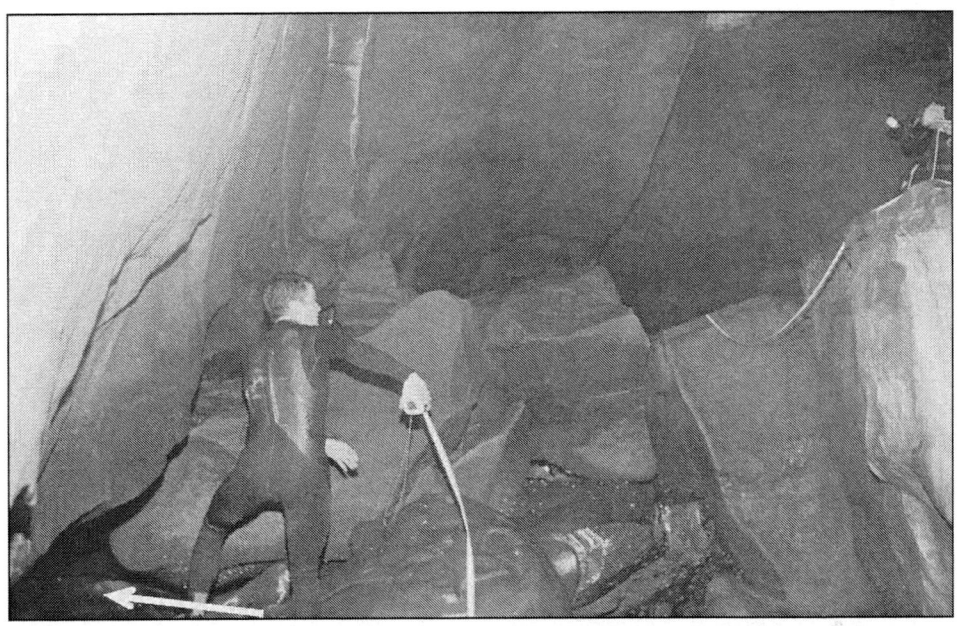

Photo 6. *Interior of Cooks Crack sea cave. Arrow points to 12- to 15-foot deep notch forming in westerly side of cave.*

Photo 7. *Interior of Cooks Crack sea cave. Note size of debris blocks.*

Photo 8. *Entrance to Cooks Crack cave, which is developing along fault trace.*

Photo 9. *Southwesterly end of Cooks Crack. Note erosion continuing along fault trace.*

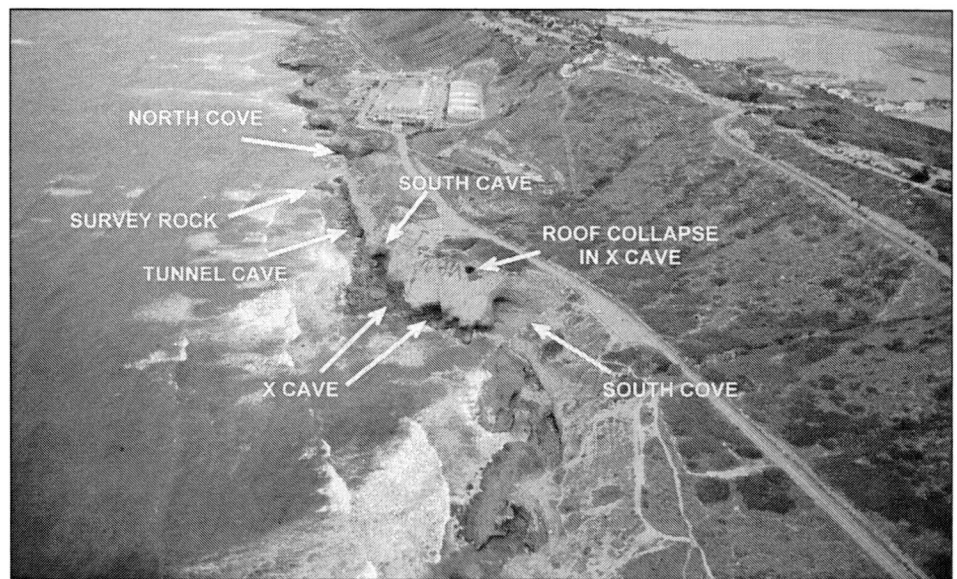

Photo 10. *PLWTP South Access Road, March 1987.*

Photo 11. *Inside X-Cave (2002).*   **Photo 12.** *Inside X-Cave (2002).*

Photo 13. *South Cave (2002). Note near-vertical fractures above cave opening.*

Photo 14. *North opening inside Tunnel Cave (2002).*

Photo 15. *South opening inside Tunnel Cave (2002).*

Photo 16. *Bootleggers caught in the Rum Runners Cave.*

Photo 17. *Rum Runners Cave accessed during low tide during City inspection-tour.*

REFERENCES

Kennedy, M.P., 1973, Sea-Cliff Erosion at Sunset Cliffs, San Diego, California Division of Mines and Geology, California Geology Magazine, p. 27-31.

Kuhn, G.G., and Shepard, F.P., 1984, Sea Cliffs, Beaches and Coastal Valleys of San Diego County, California, University of California Press, 193 pp.

Sitar, N., G.W. Clough and R.G. Bachus, 1981, Behavior of Weakly Cemented Soil Slopes under Static and Seismic Loading Conditions: U.S. Geological Survey Open-File Report 8152m 175 p.

TerraCosta Consulting Group, Inc., December 12, 2002, Coastal Bluff Stability Study, Coast Boulevard Between Prospect Street and South Casa Beach (Including La Jolla Cove and Children's Pool Beach), La Jolla, California, Project No. 2086.

TerraCosta Consulting Group, Inc., in association with Testing Engineers San Diego, Inc. April 16, 2003, 2003 Coastal Erosion Assessment Update from Sunset Cliffs Park to Torrey Pines State Beach, City of San Diego, California, Document No.: C-11542, Project No. 2162.

TerraCosta Consulting Group, Inc., October 28, 2002, Update Bluff Retreat Evaluation, South Access Road Protection Project, Point Loma Wastewater Treatment Plant, San Diego, California, Project No. 1760A.

TerraCosta Consulting Group, Inc. (formerly Group Delta-San Diego), in association with Apex Geotechnology, Inc., March 1993, City Of San Diego Coastal Erosion As-Needed Consultant Services Inventory Of Site Conditions From Sunset Cliffs Park to Cortez Place, Sites 1 through 38 Inclusive, Project No. 1487JV.

ARCHAEOLOGICAL INVESTIGATIONS OF THE SUNKEN GARDENS OF MISSION SAN LUIS REY, CALIFORNIA

Jack S. Williams
jswusu01@yahoo.com
Center for Spanish Colonial Research
San Diego, California

Anita G. Cohen-Williams
Center for Spanish Colonial Research
San Diego, California

ABSTRACT

Mission San Luis Rey was founded in 1798. It was the eighteenth Franciscan settlement created in Alta California. The outpost quickly evolved into one of the largest, and most elaborate, missions in the province. During 2003 and 2004, a team from the Center for Spanish Colonial Research undertook a major investigation of the so-called "Sunken Gardens" area of the site. A large midden with abundant colonial-era and native-tradition artifacts was identified. A wide variety of different structures were later superimposed on this trash deposit. It appears to represent the location of the contact-period native village of Tacayme. A number of features, including a series of cobble-lined acequias and pools, were created during the early years of the mission. During the decades that followed, the village habitations were relocated, and the area was reorganized. Before 1835 it included gardens, huertas (orchards), bathing and washing facilities, and an industrial zone. The project recorded a massive lime/brick kiln, a smaller ceramic kiln, tile-lined aqueducts, two tanning vats, a masonry water-filter, a fired-brick bridge, two lavanderias, a monumental stairway, a monumental gate, an adobe enclosing wall, and a number of other significant architectural features. The project staff combined archaeological and documentary evidence to create graphic images of the development and visual characteristics of the Sunken Gardens. The new investigations revealed a diverse array of features and objects that relate to the transformation of the Luiseño people.

Mission San Luis Rey is a well-known landmark that receives more than 100,000 visitors a year. The boundaries of the archaeological site have never been firmly established. The description in the 1970 nomination of the mission as a *National Historical Landmark*[1] mentions that it included at least thirty-five acres. Additional archaeological investigations expanded the site definition to include not less than the sixty-two acres in 1989.[2] Similar work by Magalousis and Kelsey revised this figure to approximately one hundred and eighty acres. However, more recent investigations have made it clear that part of the archaeological site extends into adjacent properties owned by the City of Oceanside, other government entities, and private parties.

The essay that follows summarizes some of the major findings of the 2003–2004 effort in the Sunken Gardens area of the mission (Figures 1 and 2). It is divided into sections that briefly describe the evolution of the mission, the architectural features found in the study area, and the artifacts recovered at the site. In concludes with some general observations about the significance of the findings for a broader understanding of the culture change that took place at the site.

THE EVOLUTION OF THE MISSION

Mission San Luis Rey was founded on June 13, 1798. Within a short period of time it grew into one of the most prosperous settlements in California. Father Antonio Peyri headed the missionary effort for much of the history of the site. The Franciscans pursued an ambitious development program that reached thousands of Indians. At the time of secularization in 1834, the settlement was still growing (Engelhardt 1921; Henderson 1989:4-321).

The first mission complex

Mission San Luis Rey was established at, or very close-to, the Luiseño village of Tacayme.[3] It is possible that the missionaries had already developed a detailed plan for the construction of the complex when they arrived.[4] If they did not, they appear to have quickly developed a set of objectives. Unlike most other missions, San Luis Rey was never moved.

The first phase of the mission's development called for the creation of temporary structures. Governor Diego Borica ordered that the commander of San Diego Presidio, Antonio Grajera, provide a squad of soldiers to Father Lasuén. They were to help in the creation of the initial constructions (quoted in Engelhardt 1921:8). The laborers who probably participated included a large number of neophytes from Mission San Juan Capistrano, and many local gentiles.[5]

The buildings that the newcomers initially created at San Luis Rey included huts (*jacales* or *chozas*) made with walls and roofs of grass, poles, brush, reeds, and wattle and daub. When they were available reeds (*carrizal*), tules *(Scippus lacustris)*, or brush, was interlaced to make crude mats that could be used in roofs and walls. Many of the *jacales* probably had tabular stone foundations. Some simple temporary structures were built with a wooden frame covered by thatch. The walls were built by hanging bundles of grass from horizontal poles that were attached to the wooden skeleton. The sides of the houses would have been held together by a combination of rope, twine, and leather fasteners. The floors were made from compressed earth or clay.

The first structures that were completed included a small *enramada* that was used as a chapel, and a crude residence with two rooms for the missionaries (Engelhardt 1921:9 (quoting Domingo Rivas)).[6] Some of the temporary buildings that were created may have also been of forked-pole construction.

Once temporary accommodations were complete, the construction crews quickly focused their efforts on manufacturing needed materials for more permanent structures. Before the end of July a total of more than 8,000 adobe bricks were molded, and 175 beams had been hauled to the outpost. Other building materials such as miscellaneous wooden structural elements, woven mats, metal hardware, cordage, and leather fasteners, were probably also manufactured. A cobble footing had been put in place for a structure with five rooms measuring thirty-two varas in length.

The first construction efforts were complicated by a lack of water. Antonio Peyri later reported (1827); "...*Trusting in divine providence, we established [the mission] on a broken plateau beside a marshy swamp. Filling it in little by little we managed to force the water to collect at one side of the mesa, flowing into two basins, for consumption by the inhabitants and for watering a garden*" (quoted in Kelsey and Magalousis 1991:19).[7]

The first three adobe rooms of the mission complex were completed in August (Engelhardt 1921:12).[8] By the end of December, the compound included a church, *convento* (friary or mission residence), a soldiers' barracks, and a warehouse with roofs of beams, covered with earth. Work had also been initiated on a young women's residence, or *monjerio*.[9] The walls were an adobe and a half thick (Engelhardt 1921:14; Jackson and Castillo 1995:143).

The development of a proto-urban center

Father Peyri apparently took an active part in creating a master plan for his community.[10] Unlike many of the earlier missions, San Luis Rey remained at its original site. Facilities of different sorts were added on to those that already existed in a logical manner that suggested a remarkable sense of urban planning. By avoiding labor intensive movements of existing facilities, and recycling extant buildings, walls, and other construction features, the mission rapidly developed an impressive appearance.

In 1799 construction projects included adobe rooms for the young men, a weaving room, a storeroom for wool,[11] and two granaries (Engelhardt 1921:14).[12] The next year, two rooms were added to the quadrangle and a more formal soldiers' barracks with seven apartments was completed.[13] The new structures had walls an adobe and a half thick. They had flat earthen roofs, or *azoteas* (Boyle 1968:3; Engelhardt 1921:16; Jackson and Castillo 1995:143).[14]

In 1801 a new building material made its appearance at Mission San Luis Rey. A large granary and two new rooms were completed.[15] Half of all the buildings were also improved by the addition of tile roofs (Engelhardt 1921:18).[16] Many new gleaming white buildings with dramatic red roofs soon appeared at the mission site. By 1800 the population of the mission had grown to 337 neophytes (Engelhardt 1921:16). In 1802, a new adobe church that could accommodate the growing population was completed. This structure measured 50 varas (about 137 feet) by seven varas (about 19 feet) in plan, and some six varas in height (about 16 feet). Four large chambers measuring eleven (about 30 feet) by five varas (about 14 feet) in plan, were erected.[17] The height of the structures was five varas (about 14 feet). All the remaining major buildings were also roofed with tiles (Engelhardt 1921:18).

In 1804 four granaries were built. These buildings had plans measuring 20 varas (about 55 feet) by six varas (about 16 feet). The structures' height was seven varas (about 19 feet). These new grain warehouses completed the first quadrangle. In addition to these projects, the first tanning vats[18] and a soap "boiler" were made from ladrillos (Boyle 1968:4; Engelhardt 1921:18; Jackson and Castillo 1995:143).

A new monjerio (women's dormitory) was completed in 1806. This structure had its own patio. It measured 26 varas (about 71 feet) by 16 varas (about 44 feet). A corridor connected the monjerio patio to the main quadrangle (Engelhardt 1921:19; Jackson and Castillo 1995:143; Kelsey 1993:13). This structure probably stood on the northern side of the site.[19]

In 1808 a number of other modifications were made to the mission quadrangle. Two rows of adobe building were raised to the height of the church. Two corrals were created that abutted the exterior of the compound (Engelhardt 1921:19; Jackson and Castillo 1995:143). These structures appear to have been built to the east of the quadrangle.

The year 1810 produced mention of the creation of two walls to enclose a vineyard. The area was fully walled two years later (Engelhardt 1021:20, 23). This feature may well have been included in the Sunken Gardens, or an adjacent area.

In 1811 work began on the massive neoclassical church that still dominates the mission (Engelhardt 1921:23). This was the most ambitious structure created at the mission. The design and other characteristics of the building point to the employment of a master builder. This man was José Antonio Ramírez, a master carpenter and stone mason from Mexico.[20] The work effort was not concluded until 1815 (Jackson and Castillo 1995:143; Webb 1952:130).

Construction efforts probably stalled late in 1812. A terrible earthquake that occurred on December 8, caused massive damage at San Luis Rey's northern mission neighbor, San Juan Capistrano (Engelhardt 1920:160). Despite these set backs, by the end of 1813 a new row of rooms and new interior and exterior corridors (porticos)[21] were added to the main quadrangle (Engelhardt 1921:23; Jackson and Castillo 1995:143). Work on the church, and some unidentified structures continued the next year. The structure included two bell towers, candelabras, three major altars, a massive nave, a sacristy, a baptistery, and an elaborate choir loft (Tac 1952:95). On October 15, 1815, the magnificent house of worship was finally dedicated (Engelhardt 1921:35).

With the completion of the great church and surrounding brick porticos, the mission quadrangle attained a spectacular appearance. During the years that followed, the settlement continued to witness a number of other major construction efforts. However, they pale by comparison to the achievements of 1798-1815.[22]

The later development of the mission

Although the main quadrangle had been completed, other mission areas were still incomplete. In a letter from Antonio Peyri to Governor Sola written in 1817, he noted that the new church needed repairs. Specifically, a soldier-builder Salvador Bejár, needed the help of more skilled carpenters and blacksmith. He further stated that the builders were having difficulty setting tiles (quoted in Engelhardt 1921:36; Schuetz 1994:159).[23] During the next two decades, important maintenance projects had to be completed. A number of other buildings are noted that were probably constructed during the latter part of the mission era. These structures included residences, workshops, grist mills, and store rooms.[24] Important changes and additions were also made to existing buildings, including the church and the main quadrangle[25] (Anonymous 1865; Cleal 1854; Duhaut Cilly 1929; Duel 1921; Engelhardt 1921:38-39, 56, 67; Hancock 1860; Kelsey 1993:12; Robinson1829; Schuetz 1994:159; Tac 1952 - Figure 3).

The reuse of the mission complex

After secularization, the mission's sprawling buildings were shared by priests, resident Indians, and a diverse array of civil administrators (Engelhardt 1921:147-55; Hartnell 2004:60-62; Salomon 2004:95). The principal lands of the former mission were eventually transferred to public ownership by Governor Pío Pico at the end of Mexican rule. The United States' Army occupied the mission from 1847-1849, and again in 1850-1852. During this period they used existing buildings and set up tents (Hart 1967:66; 2001). After the army departed in 1865, treasure hunters ravaged the buildings, especially the church (Weber 2001:91). The looters were eventually joined by local residents who salvaged tile and other reusable materials (O'Keefe quoted in Healy 1992:17). Eventually, the structures were abandoned, and the mission property returned to the Catholic Church. The Franciscans returned in 1892. Since then, large portions of the mission have been rebuilt, and numerous new structures constructed at the site.

THE ARCHITECTURAL FEATURES OF THE SUNKEN GARDENS

The Sunken Gardens have a complex set of architectural features that date from the colonial era. The area played a large part in the life of the mission.[26] The archaeological investigations documented a number of important features of the mission landscape (see Table 1, figures 4–24).

Perhaps the single most important architectural discovery of the project was the identification of a set of two tanning vats. These structures had previously been partially exposed and identified as a tile floor. The present project re-exposed the remains, which consisted of the lower portion of a large, two-chambered structure made out of ladrillos. The annual report of 1804 lists the creations of *"two capacious tanks of brick"* at the mission (Engelhardt 1921:18).

The feature consists of the remains of a floor made up of a number of layers of bricks, and the lower courses of ladrillo walls. The floor areas had abundant traces of white powder that appeared to be quicklime. Some of the material had a yellow discoloration. In contrast with the floors, the walls were set in mud mortar. They did not rest on any kind of cobble foundation, and were not set as deep as the floors. A hole had been dug through the floor of the southern chamber at an earlier date. This allowed us to examine the multi-layered structure of the covering. About 75% of the upper surface of the feature was exposed through excavation. Our identification of the function for the structure is based on the general shape of the chambers, the availability of nearby water, and the abundance of traces of a pale yellow powder (probably traces of the tanning solution) in the several layers of bricks that formed the floors.

Another important discovery was made in connection with the large kiln that has been variously identified as a brick or lime kiln. Excavations in the throat of this structure identified both tile wasters and limestone clinkers. Thus, it seems reasonable to suppose that the kiln was used to make both quicklime and tile during the colonial era.

The architectural remains in the Sunken Gardens suggest that a wide variety of tasks took place there in colonial times. The tract was used for agriculture, recreation, some domestic activities, and industry. Specific work areas included spaces devoted to bathing, recreation, food production, hide processing, brick production, quick-lime production, and cloth washing.

ARTIFACTS RECOVERED BY THE PROJECT

Nearly 11,000 artifacts were recovered from the Sunken Gardens by the 2003 excavations. They included building materials, ceramics, glass, bones, seashells, and limestone. The previous investigation of the area by Soto produced an even more diverse collection. He reported recovering native pottery, pottery disks, two native smoking pipes, bone gambling sticks, gaming stones, an arrow shaft straightener, metates, mortars, flint knives, arrowheads, spearheads, hundreds of Indian and glass beads, Chinese pottery, Mexican pottery, coins, musket balls, a religious medal, a musket trigger guard, iron spikes, nails, sea shells, and charcoal (1960, 1961).[27] The limited test excavations conducted in the Sunken Gardens produced a less elaborate, but fundamentally similar, collection of mission-era artifacts. In addition to the 9,778 colonial artifacts, 1,133 objects dating from the period after 1900 were found.

A total of 7,537 fragments of imported and locally produced ceramic objects were recovered from the Sunken Gardens (see the table that follows). A significant portion of the collection (3,699 pieces) is represented by tile. This architectural debris is associated with the masonry constructions found in the area. Some of the fragments were unintended byproducts of the tile manufacturing process. They were probably created at the kilns located within the study area. The remaining materials represent sherds of ceramic vessels that were used for cooking, food preparation, food serving, and storage (see Table 2).

A variety of different types of stone artifacts have been found at Mission San Luis Rey. These include chipped stone and ground stone objects. In addition to these categories, the project also recovered a number of fragments of limestone that were by products of industrial processes.[28] Table 3 summarizes the major groups, and distribution, of stone objects from the Sunken Gardens.

Copper and copper alloys are generally well-preserved at the mission. They can be distinguished from other metals on the basis of greenish, and greenish-blue, colored oxides. A single fragment of thin copper, or copper alloy, was collected. The specimen recovered probably represents a fragment of a colonial vessel. Common forms for these containers include *tazas* (cups), dippers, *calderetas* (small cauldrons), *jarros de latón para agua* (brass water jars), and *chocolateros* (chocolate preparation jars). The most popular form appears to have been cauldrons. Most of these vessels were made from relatively thin sheet metal, although some examples appear to have been created as one-piece castings. Many of the copper and brass objects that found their way to the frontier were manufactured as a by-product of the thriving silver industry in Central New Spain (Barrett 1987; Di Peso 1974(3):947; Webb 1952:144).

The bones of animals represent one of the most common kinds of remains recovered from Spanish colonial sites. This pattern was visible in the artifacts recovered from the Sunken Gardens. A total of 490 specimens were unearthed. These bones were widely distributed throughout the study area. They were present in nearly all the deposits that were investigated. Most of the materials that were encountered represented large land mammals that had been consumed as food. Smaller amounts of bones from other animals, including wild species, were also collected. These creatures had entered the archaeological record as a result of both human, and natural, agencies. The zooarchaeological remains recovered are summarized in Table 4.

Many of the large land mammal bone fragments recovered from the Sunken Gardens provide evidence of the neophytes' butchering practices. The tools and methods used in the butchering process can be identified by the forms of breakage seen at the ends of fragments (such as abrupt breaks produced by chopping), as well as less obvious marks on the bone surfaces. Thin cuts produced by knives appear as V-shaped lines of shallow to moderate depth. Cleavers, axes, and chisels, leave behind abrupt break lines (over three mm deep). Saws produce lines that have distinctive parallel side notches. The overall pattern seen in the butchering marks at San Luis Rey is consistent with those noted at other Spanish Colonial sites, such as San Diego Presidio (Cheever 1982) and Tubac Presidio (Arizona - Hewitt 1975).

A total of 269 marine shells, and shell fragments, were recovered from the Sunken Gardens. The specimens represent a diverse array of types. Nearly all of the varieties that have been observed could be collected from nearby coastal waters. The shells, and shell fragments, were widely distributed over the entire study area (see Table 5). Marine shells came to San Luis Rey through a variety of mechanisms. Most of the specimens recovered probably represent food waste. Some shell beads, such as the one made from *Olivella biplicata*, were imported. Locally acquired materials were also used to make jewelry at the site. During the early days, significant amounts of marine shell were brought to the mission to be burned to make quick lime. This substance was a key ingredient in plaster and cement mortar. Eventually sea shells were replaced by limestone.

A significant number of glass fragments were found in the Sunken Gardens. Most of these items can be dated to the twentieth century. No reconstructable vessels or glass beads were recovered. A total of 13 sherds of so-called *"black glass,"* were unearthed. These fragments were dark-olive in color. The sherds were too small to suggest any specific vessel forms. When compared to later historic eras in California, glassware is relatively rare in colonial sites in California. The scarcity of glassware may have resulted from the extensive use of ceramic vessels for similar purposes.

A number of seeds and other kinds of miscellaneous organic materials were recovered from the Sunken Gardens. The eleven seeds were not carbonized. They appear to have been dropped by modern trees. Sixty small carbon nodules were found. They probably represent the remains of materials used as fuel for heating, cooking, and industrial processes. The carbon may also have been the products of natural brush fires. A single piece of carbonized wood may be explained by similar site formation processes.

Table 6 provides a summary description of the quantities and character of the artifacts recovered from the Sunken Gardens in 2003. The overall pattern of the relative frequencies of artifacts, sheds light on some important aspects of the use of the Sunken Gardens. Over half of the collection consisted of building materials (principally tile, soft cement, and similar mortar fragments). No residential structures that used these elements are thought to have existed in the study area. The building remains are almost certainly connected with the industrial, agricultural, and communal facilities, that are represented by the features noted elsewhere in this essay.

Another important characteristic of the collection is the preponderance of kitchen-related objects. Nearly half of the collection (46.9%) is represented by this category of artifacts. Clearly, food preparation was a common activity in the Sunken Gardens. The distribution of objects in various strata that were observed, as well as the types of datable ceramics that were encountered, make it clear that the cooking was taking place during the mission era (rather than in more recent, or pre-contact times). The array of associated artifacts also suggest a household, rather than communal, form of food preparation. No documents specifically mention that the study area was used for habitation, but the archaeological record clearly indicates that it probably served this function. These findings were independently confirmed by Soto, who also found abundant domestic artifacts in the area (1961:34). The presence of colonial artifacts in the lowest strata, combined with information about the later use of the study area (which did not include any residential activities), suggests that this is the location of a native mission era occupation.

When did this occupation take place? Prior to the creation of the water system that brought the precious liquid from the springs to the east in 1808, the Sunken Gardens area and its immediate proximity would have been the most convenient place to locate the neophyte village (Engelhardt 1921:19). The end of this occupation appears to have occurred less than a decade after the settlement was founded, when the area was converted to an agricultural and industrial zone.[29] Thus, it appears that the Sunken Gardens contain the remains of the first mission Indian village, dating to circa 1798-1808.

CONCLUSIONS

The remains recovered by the 2003-2004 investigations provide insights into the transformation of the Luiseño culture during the mission era. The remains found in the study area reflect diverse cultural origins. The presence of different technological traditions points to a complex blending of civilizations that produced a unique frontier people.

One of our major research objectives of the project was to determine the cultural sources of various items found in the study area. The excavations unearthed materials of European, local Native American, Mesoamerican, and African origins. The European cultural elements include the introduction of the use of fired bricks, the control of water using lined channels, the use of paved walkways, and the use of Old World cooking wares (such as galeraware) and serving wares (including Chinese porcelain and maiolica). The importance of the Old World traditions can also be seen in food ways. New species that appeared in Luiseño meals included cattle, sheep, possibly horse, and chickens.

Native traditions derived artifacts are also abundant. Specific items of Luiseño origins include ground stone objects, chipped stone objects, jewelry (represented by a single shell bead), and native tradition pottery. In regards to Mesoamerican cultural elements, it is clear that the broader tradition of Northwestern New Spain has many items that have origins in ancient Mexico. In the case of Alta California as a whole it is obvious that many items of Mesoamerican origin also made their way into the Franciscan directed program of culture change. The most obvious of these items may be domesticated corn, certain kinds of beans, and turkeys. At least one of the classes of fine, decorated ceramics (*Bruñida de Tonalá* ware) found elsewhere at Mission San Luis Rey, can also be traced to pre-Columbian origins. The presence of these objects suggests that models of frontier culture change need to incorporate non-local Indian, as well as Old World cultural influences. A specific item of note that has Mesoamerican connections is the rim of a ceramic comal (griddle). This object reflects the introduction of a new form to a traditional material.

African cultural elements are scarcer, or are less obvious. The most probable link can be drawn to butchering styles that are visible in the zooarchaeological remains. The systematic reduction of bones for marrow, and their subsequent boiling are customs shared by northern and western African peoples, as well as the people of New Spain, and the northwestern frontier. A more subtle inspiration that arrived in Spain, via North Africa, was the focus on washing and cleanliness in worship. This cultural tradition is particularly obvious in Islam. By contrast, Northern Europeans of the later medieval period placed far less emphasis on being clean.

In what ways do the remains found in the Sunken Gardens reflect culture change? The missionaries of Alta California arrived on the frontier with a definite mental template as to the changes that they anticipated making in the Indian community. From an archaeological perspective, the most obvious manifestations of these changes are likely to be seen in the areas of architecture and economics (especially foodways). At San Luis Rey Mission the changes can be seen through a number of technologies, and food resources.

In regards to the settlement systems, it is clear that if at all possible, the missionaries attempted to draw the population together into compact towns (*reducciones*) that were consistent with European notions of towns. The key elements of their ideas can be traced to certain broad concepts involving Iberian town planning. Villas and pueblos were constructed from permanent materials. They were laid out around a square or rectangular central plaza (a *plaza de armas*). On one side of the plaza a church was constructed with abutting residence building and offices. On the opposite side government buildings (often including the *cabildo* (town council building and government headquarters), *carcel* (jail and police station) and a municipal granary) would be erected. On the remaining sides of the plaza would be built a combination of workshops, warehouses, and residences. Traditionally, the more valuable house lots (*solares*) were located on the plaza. Gardening plots and fields were developed in areas that did not face the plaza. Larger industrial areas were also constructed at some distance. This basic plan was often adapted to the special needs dictated by topography, or by defense concerns.

The limitations imposed by the scarcity of building materials and technical expertise on the frontier usually prevented the immediate realization of the missionaries' visions. During the earliest days, temporary materials, such as cloth, wood, and mud, were used to establish provisional facilities. Over time, the crude structures were supplanted by more permanent buildings of adobe and undressed masonry. With prosperity and time these edifices give way to those of cement, brick, and cut stone.

In the case of Mission San Luis Rey, the community almost immediately adopted the features of what I have called a proto-urban phase. Between 1798 and 1810, temporary materials were replaced by massive adobe and tile constructions. The rapid advance of the building arts at the settlement is undoubtedly related to its population growth and economic prosperity. In any event, the creation of the elaborate architectural features of the Sunken Garden point to an unprecedented transformation that has broad implications for the restructuring of the Luiseños way of life. New ideas about washing clothing, cleaning polluted water, building, and processing leather, were introduced. At the same time, the fact that the native residents of Mission San Luis Rey did not give up their traditional housing elsewhere at the settlement also points to an unusual flexibility on the part of the Franciscans. In that sense, it remains clear that the Luiseños were adopting elements that they found useful. The changes that were going on were probably more additive than replacing.

A few of the artifacts recovered in the Sunken Gardens have direct significance for the mixing of cultures. A chipped ladrillo biface was recovered.

It represents a particularly interesting example of hybridization. Cheek's work at Mission San Xavier del Bac provided a few examples of similar re-uses of fired bricks (1974). The impact of domesticated animals on Luiseño diet was both dramatic and obvious. The first cattle and sheep that arrived at the mission were contributed by the Franciscan communities at Santa Barbara, San Diego, San Juan Capistrano, and San Gabriel. The initial herds counted more than 310 cattle and 508 sheep (Engelhardt 1921:14). They rapidly increased in number and provided a reliable source for food, raw materials for industry, and a valuable item useful in commerce. Horses, mules, burros, and oxen, also provided valuable transportation.

The apparent preference for domesticated animals by the Luiseños, particularly for beef, reflects their adoption of Iberian colonial dietary practices (Hewitt 1975:206).[30] Nineteenth-century visitors to California also stressed that northern settlers preferred beef to all other foods (Cheever 1983:39, 43; Webb 1952:99). Sheep and goats provided an important secondary resource (Engelhardt 1920:43; Cheever 1983:38).

The introduction of some new technologies appears to have resulted in a decline in the importance of some artifacts. The quantity and quality of chipped stone objects apparently diminished as they were replaced by more effective metal tools.[31] The Luiseños had limited access to useful raw materials. It is therefore not too surprising that they were open to the adoption of alternative technologies. The chipped stone industries observed in San Diego County colonial sites generally show a decline when they are compared to late prehistoric sites.[32] This is probably a direct consequence of the introduction of metal tools that served similar purposes.[33]

The enthusiasm shown by some natives for items of Old World origins did not lead to the complete abandonment of older traditions. The archaeological record makes it clear that the mission inhabitants were involved in a selection process where items that were deemed to be useful were adopted and used alongside traditional technologies. The presence of deer bones, fish bones, and marine shells, suggest that hunting,[34] gathering, and fishing, were still of some economic importance to Luiseño neophytes. Traditional native pottery, albeit with some modifications, continued to play an important part in Luiseño life. In this sense, it remains clear that the culture change seen in technology at Mission San Luis Rey was sometimes additive, rather than replacing.

The changes seen at Mission San Luis Rey need to be understood in terms of a broader context. Native peoples were clearly being transformed in a political and social sense into elements of a colonial society. The people anthropologists have labeled Luiseño had conceived of themselves in earlier periods in terms of their village communities. Although they shared certain material and linguistic traditions, they remain politically autonomous. The creation of the mission integrated these people into a larger political identity (as vassals of the Kingdom of Spain, and later, citizens of the Republic of Mexico). This integration process allowed for cultural pluralism, in the sense that they were recognized as both Native Americans, and people of a particular regional area. For Spaniards and Mexicans, these people were associated with the urban center of Mission San Luis Rey. In that sense, the additive and replacing technologies seen at the mission site show the material aspects of the formation of a new Luiseño culture identity that was both uniquely a kind of Colonial American phenomena, and a continuation of its pre-contact predecessors.

Researchers who are interested in more details about the project should contact Jack S. Williams (jswusu01@yahoo.com), the principal investigator of the project. A detailed report on the project that has been submitted to the mission, which includes 483 pages of text and illustrations, will soon be available in PDF format (Williams 2004).

Adapted from: Architecture, Physical Environment and Society in Alta California. Proceedings of the 22nd Annual Conference of the California Mission Studies Association, Mission San Fernando Rey de España. Rose Marie Beebe and Robert M. Senkewicz, eds. February 18–20, 2005.

NOTES

1. USI/COMOS (United States Committee, International Council on Monuments and Sites) *Preliminary Inventory of Spanish Colonial Resources Associated with National Park Service Sites and National Historic Landmarks, 1987 second edition* (United States Department of the Interior: Washington, 1989), pp. 4-321 to 4-327.

2. The site definition was provided by Debra A. Dominici. See Magalousis and Kelsey 1992.

3. When Iberians arrived in 1769, they found the native peoples living in two villages on the edges of large pools of water in the nearby main channel of the river (Kelsey 1993:2). The village of Tacayme may have developed between 1769 and 1798.

4. As suggested by Kelsey (1993:11).

5. See also, Bancroft (1884:563-64).

6. The rooms measured approximately ten yards in length and five in width.

7. The original document was located and re-translated by Kelsey and Magalousis (1991:19). Engelhardt offered a somewhat more confusing translation "... *there is no running water nor any spring* [at the mission proper] ... *however, we established a mission on a mesa situated near a marsh, the water of which by hauling earth we succeeded little by little in forcing up so that it could be reached. By means of two dams the water was then collected so that it sufficed for the assembled Indians and for irrigating a garden."* The marshy area is believed to have included the lower study area.

8. The French visitor of 1827, Auguste Duhaut Cilly later quoted Father Peyri's remembrance of what happened; *"The Father [Peyri] related how he arrived at 4:00 in the afternoon, on June 13, 1798, on this then still uninhabited plain, accompanied by the comandante of San Diego, a troop of soldiers, and workmen.* "Our first care was," he said, "to erect some kind of a cabin after the fashion of the savages of this region, since it was to serve as a shelter until the mission should be built; but on the following morning, before excavating for the foundation, we raised in the open air an altar. Here under the canopy of the sky was offered up for the first time in this valley the holy Sacrifice to the Eternal, which thereafter culminated in such great blessings" (quoted in Engelhardt 1921:12). The full text of the related narrative can be found in Duhaut Cilly (1929:131–165).

9. Completed in 1799. A detailed review of the functions of the monjerio is provided by Webb (1952:27-28). She notes that the apartments were inevitably set in the *"least disturbed part of the square"* (1952:104). The origins of the monjerio can be traced to Medieval times, when noble women were placed in seclusion in a *gynecaeum (gynecaeum)* to protect them from insults and prevent misbehavior (Duby 1988:69, 77–83; Regnier-Bohler 1988:344; Roncière 1988:214–215).

10. Duhaut Cilly later noted; *"The buildings were planned on a grand scale after the idea of the Father [Peyri]. He himself and alone could have superintended the construction* ..."(quoted in Engelhardt 1921:55).

11. These rooms may be referred to as facilities for large, and small, looms in Tac's 1835 memoir (1952:96). Two rooms for looms, a room for carders, and a room for spinning are also shown in the west and northern wings on Mofras' map (1845). Webb reviews the history and methods employed by weavers in California (1952:207–216).

12. These may have been the storerooms noted by Robinson in 1829 (quoted in Engelhardt 1921:67). Common features in mission granaries are described by Webb (1952:58).

13. The barracks are shown in several illustrations (Anonymous n.d.; Bartlett 1852; Duhaut Cilly 1829; Miller 1856; Powell 1850, Powell 1850a; Robinson 1829). Plans are provided by an anonymous author (1865), Cleal (1854), and Hancock (1860). An earthquake damaged the mission to an unclear extent the same year (1800, Engelhardt 1920:154).

14. This is apparently the same structure that would serve the mission community for the rest of its history. It was mentioned in Tac's circa 1835 memoir (1952:96). In 1827 Duhaut Cilly notes; "*In addition to the immense main building I have just described (the extant main quadrangle), there are two others much smaller, one of which is given up to the mayordomos, the other to the mission guard composed of a sergeant and eleven soldiers* (the extant soldiers' barrack complex). *This latter building has a flat roof and a dungeon with barbicans and loopholes*" (1929:228).

15. Boyle claims four rooms (1968:3).

16. The re-roofing project was apparently completed in 1802 (Schuetz 1994:159). San Luis Rey was one of the few locations in California where roofing was created using ladrillos. A paved masonry terrace was created above the colonnades of the main quadrangle (Duhaut Cilly 1827 quoted in Engelhardt 1921:56; Miller 1985:55). These constructions were similar to other kinds of azoteas, except the upper surface was paved with a combination of tiles set in concrete.

17. Boyle suggests that they were granaries (1968:3).

18. These may be represented by feature 24 in the Sunken Gardens.

19. This was probably the girls' dormitory noted by Robinson in 1829 (quoted in Engelhardt 1921:67). See also Tac (1952:96). The anonymous plan of 1865, Cleal's 1854, Hancock's 1860, and Oak's 1874, plans, show extensions on the north side of the quadrangle that may represent this structure. Kelsey was convinced that the monjerio stood at the northwestern corner of the quadrangle (1993:40-41). However, Duel's map shows that the only fully-enclosed space was near the square formed on the middle of the north side, of the north wing (1921). No monjerios have survived in California. A structure of this type was "restored" at La Purísima in 1949, but it has unclear connections with the historic use of the building or even the site area (Henderson 1989:4-302).

20. Ramírez came from Zapotlán Grande in Jalisco. He was born a criollo around 1762. Ramírez came to California in 1792. He also worked on the buildings at Monterey, Carmel, San Juan Capistrano, La Purísima, Santa Barbara, Los Angeles, and San Gabriel. Ramírez died in 1827 and was interred in the Pueblo of Los Angeles (Schuetz 1994:85-87, 159).

21. Eventually, the main building would have a facade lined by thirty-two square pillars (Duhaut Cilly quoted in Engelhardt 1921:56). Boyle suggests that the row was added to the exterior of the quadrangle (1968:4).

22. The last annual report that included any information on construction was submitted in 1824 (Engelhardt 1921:38).

23. Schuetz identifies the man as Salvador Bejár. He was a criollo born in Tepic, Nayarit, who lived between circa 1767 and 1824. Bejár served in California primarily as a soldier. He also worked at Santa Cruz, Santa Barbara, and the presidios of Monterey and San Diego. He was buried at San Gabriel (Schuetz 1994:55-57).

24. Tac's memoir of 1835 mention workshops for masons, an area with a wine press, and a soap house (1952:95). Mofras' 1841 plan shows rooms for the carpenters and joiners, coopers, tailors, leatherworkers (shoemakers, saddlers, and harness-makers), smithies, and locksmiths, ironworkers (1845). Kelsey suggests that other workshops were built in a linear extension of the eastern wing that stood to the north of the church (1993:14).

25. The main quadrangle was gradually expanded through additions that were built along the exterior row of rooms. These extensions incorporated courtyards and enclosing walls. In all cases, the basic grid pattern of the settlement was continued. By 1817, some form of hospital had been erected at San Luis Rey (Schuetz 1994:159). A decade later, Duhaut Cilly noted the existence of an infirmary and private chapel in the main mission compound (quoted in Engelhardt 1921:56–57; and Webb 1952:287). The structure was divided into three wards, a chapel, and a pharmacy. Two rooms with privies flanked the complex. The hospital's children, male, and female, patients were assigned to respective wards (See Osio 1996:124–125).

26. Tac's memoir of circa 1835 provides an unusually detailed description of the main mission orchard. It includes information about several structures that were apparently created after 1821 (Tac 1952:96–97); "*Towards the south there is a very big kitchen garden with a pasture to the side. We said that the mission was placed on a hillock. Below the hillock there is an ever-flowing fountain from which the neophytes and the missionaries bring water to drink. They made two fountains before the gate of the garden, and between them a stairway to go up and down and which is all made of bricks. The entering gate has three thick timbers in the middle. One of them driven into the earth reaches high above the wall, the other two more or less fastened to it making a cross of all parts . . . and the water carrier wishing to pass pushes a timber, and the two turn, and in this way he passes with ease, raising the pitcher above his burdened shoulders stronger than those of asses themselves. The stairway is so very high, that one cannot ascend it by the same trip, and it is necessary to rest in the middle. It happens many time that they get tired in vain (as is said), because they arrive at the gate and wish to pass through it with haste, the pitcher is broken, and the return to the house without water or pitcher, dripping with water. The timbers were placed in order not to let in the bulls and horses, spirited when there is bullfighting, though they come in often and frighten the old women who wash their clothes here. Beyond the two fountains is the gate to the orchard. The water from the two fountains passes through a little door, running towards the west as in a ditch, and irrigates another garden almost a league from the Mission. The garden is extensive, full of fruit trees, pears, apples or perones, as the Mexicans say, peaches, quinces, pears, sweet pomegranates, figs, watermelons, melons, vegetables, cabbages, lettuces, radishes, mints, parsley and others I do not remember. The pears, apples, peaches, quinces, pomegranates, watermelons, and melons are for the neophytes, the others that remain, for the missionary. The gardener must bring something each day. None of the neophytes can go to the garden or enter to gather*

fruit. But if he wants some he asks the missionary who immediately will give him what he wants, for the missionary is their father. The neophyte might encounter the gardener walking and cutting the fruits, who then follows him to punish him, until he leaves the walls of the garden, jumping as they know how (like deer in the mountains). Once a neophyte entered the garden without knowing the gardener was there, and was very hungry, he climbed a tree. Here he began to eat with all haste a large ripe fig. Not by bits but by whole he let it go down his throat, and the fig choked him. He then began to be frightened, until he cried out like a crow and swallowed it. The gardener hearing the voice of the crow, with his Indian eyes then found the crow that from that fear was not eating more. He said to him, "I see you, a crow without wings. Now I will wound you with my arrows." Then the neophyte with all haste fled from the garden. Duhaut Cilly also briefly described the gardens as they appeared in 1827 (quoted in and Webb 1952:76); "*Two well-planted gardens furnish an abundance of vegetables and fruits of all kinds. The large comfortable stairway by which one descends into the one to the southeast [the former marsh], reminds me of the orangery at Versailles; not that their material was as valuable, or the architecture as splendid; but there was some relation in the arrangement, number, and dimensions of the steps. At the bottom of the stairs are two fine lavers in stucco; one of them is a pond where the Indian women bathe every morning; and the other is used every Saturday for washing clothes. Some of the water is afterward distributed into the garden, where many channels maintain a permanent moisture and coolness.*" The description contains a number of apparently confusing, or erroneous, facts. Both the lavanderias were used for washing clothing. The ponds used for bathing were above the level of these structures (considered here as part of the fountains). Engelhardt offered a similar alternate translation (1921:57); "*The vast gardens and orchards with numerous fruit trees are well cultivated and supply abundant vegetables and fruits of all kinds. The sight of the wide and convenient stairway that leads to the orchard to the southeast put me in mind of the citrus fruit conservatory at Versailles, not because the material was as precious and the architecture as splendid, but because there was some similarity in the disposition, number, and dimension of the steps. At the foot of the stairway are seen two beautiful lavatories in stucco. One of them is a pond in which the Indians bathes every morning; the other serves for washing the linens every Saturday. A part of the waste water runs off into the garden in which numerous conduits preserve continual humidity and freshness.*"

27. The location of these artifacts, and any additional study notes, has not been established. Relevant repositories that were examined included Mission Santa Barbara and Mission San Luis Rey. According to Soto, the identification of artifacts was made by M. R. Harrington of the Southwest Museum (1961:42). Soto reasonably concluded that the area included some kind of midden.

28. A number of pieces of limestone were unearthed in the Sunken Gardens that are probably associated with the kilns located at the eastern end of the study area. These include seven misfired specimens. Normally, the heat in the kiln reduces the

limestone to a powdery mass. However, some limestone has mineral inclusions that cannot be easily calcified (Costello 1977:27). When the quicklime is removed from the kiln, these stones, called *clinkers,* are discarded. They are commonly associated with kilns used to process limestone (Concentrations of similar materials were found at the monumental lime kiln at Mission San Juan Capistrano — Williams 1995). As a result of being heated to extreme temperatures, many clinkers exhibit a grey discoloration. Other specimens exhibit pitting and cracking that is caused by thermal stress. One other notable specimen in the Sunken Garden collection consists of a shaped lump of limestone. Quarried stone was often reduced in size and shape to facilitate its transport, and the complete calcification of the raw materials in the kiln during firing.

29. After the creation of the new aqueducts that brought domestic water to the upper part of the mesa, work apparently shifted to an effort to create regular adobe housing for the neophytes living at the mission. Tac's circa 1835 memoir notes that the daily domestic water used by the Indians was collected from the fountains. He also states that the garden had a wall and that neophytes were not allowed to enter the area without the permission of the caretaker (1952:97). At some point before 1827, the adobe structures were abandoned. It seems reasonable to suppose that the structural elements of the houses would have been removed and reused in other mission construction projects. The new village was made up of more traditional brush housing, and appears to have been dispersed over a large area (Duhaut Cilly 1827 quoted in Engelhardt 1921:57–59; Miller 1985:55; Osio 1996:1245–125).

30. Similar patterns have been noted at Mission Santa Inés (Walker and Davidson 1989), Mission San Buenaventura (Romani and Toren 1975), Mission San Antonio (Hoover and Costello 1985), Mission San Juan Bautista (Farris 1991), and Mission Santa Cruz (Allen 1998).

31. It is interesting to note that Temeku, a contact and mission era site located in the northern Luiseño territory, had abundant chipped stone objects, including projectile points (McCown 1955). Boyle also noted numerous chipped stone objects in prehistoric sites located in the area surrounding the mission (1968:20 — including projectile points, knives, and scrapers). The combined total of the stone objects (52 items) from our work, is especially unimpressive when compared to some other classes of native artifacts, such as ceramics. The excavations at the Luiseño site of Temeku produced a projectile point made from bottle glass (McCown 1955:15). Similar trends were noted by Deetz at Mission La Purísima (1962).

32. Allen notes that this may not have been the case in Northern California (1998:77).

33. Stone tools of this type were made and used by both the Latino and Indian settlers in the region (Williams and Cohen-Williams 1987).

34. Tac describes the persistence of hunting during mission times (1952:100).

Table 1. *Colonial features of the sunken gardens, 2003.*

FEAT. NUMBER/DESCRIPTION	CONDITION	IN STRATA	OTHER COMMENTS
01 - Gateway Arch	restored	surface	previously excavated
02 - Stairs	restored	surface	previously excavated
03 - Northern Garden Wall	restored	surface	previously excavated
04 - Eastern Stair Wall	restored	surface	previously excavated
05 - Patio between lavanderias	restored?	surface	feature may represent a modern creation
06 - Western Lavandería	restored	surface	previously excavated
07 - Eastern Lavandería	restored	surface	previously excavated
08 - Bridge	restored	surface	previously excavated
09 - Acequia segment A	restored	surface	previously excavated
10 - Acequia segment B	restored	surface	previously excavated
11 - Acequia segment C	restored	surface	previously excavated
12 - Filter House	restored ?	surface	previously excavated
13 - Western Fountains	restored	surface	previously excavated
14 - Eastern Fountains	restored	surface	previously excavated
15 - small arch	restored	surface	previously excavated
16 - large kiln	stabilized	surface	previously excavated
17 - small kiln	restored	surface	previously excavated
18 - wall footing	stabilized	surface	previously excavated
19 - cobble acequia	stabilized	surface	previously excavated
20 - walkway	stabilized	surface	previously excavated
23 - footing	intact	surface	previously excavated
24 - tanning vats (structure 1)	intact	3	probably ladrillo tanks used as tanning or mixing vats; previously excavated
25 - historic gate	restored	surface	previously excavated
30 - stone platform	intact	surface	
31 - western stair wall	not visible	surface?	covered by plants

Notes: Strata 3 represents use surfaces at the site.

Table 2. *Summary of types of ceramic objects recovered from the sunken gardens, 2003.*

TRADITION	TYPES	#	NOTES
Ibero-American tradition	Lead-glazed ware	1	probably imported from northwestern Mexico; possibly manufactured in Alta California
	Maiolica	10	imported from Puebla in Central New Spain
Asian tradition	Chinese Porcelain	7	imported from Asia via Canton and Manila
British tradition	European Porcelain	3	imported from Europe, principally Great Britain
	Transferware	6	
	Cottage ware	3	
	Pearlware	1	
	Ironstone	74	
Native tradition	Plainware	3248	probably manufactured at the site
	Redware	123	probably manufactured at the site
	Buffware	358	imported from native peoples living to the east
Tile	tejas (roof tile)	271	manufactured at the site
	ladrillos (floor tiles)	203	
	wasters (over-fired tiles)	115	
	unidentified	3110	
TOTAL		7533	

Note: in addition to the specimens listed in the table above, four pieces of Terracotta were collected. Ceramics with Terracotta-like characteristics were produced during the colonial era. However, the items recovered probably date to the modern era.

Table 3. *Summary of types of stone objects recovered from the sunken gardens, 2003.*

TYPE	#	NOTES
Chipped stone objects	41	includes flakes, cores, and bifaces of a number of different types of locally available materials, including quartz, quartzite, rhyolite, and basalt; all items measuring twice as long as they are wide, or thick, are considered to be flakes; chipped stone tools were generally used for cutting and piercing.
Ground stone objects	3	includes two granitic manos and a possible basalt polishing stone; groundstone objects are generally associated with food processing
Limestone industrial debris	8	includes 7 limestone clinkers and a single shaped lump; all the items are associated with quick lime production
TOTAL	52	

Table 4. *Summary of types of zooarchaeological remains recovered from the sunken garden, 2003.*

TYPES	#	NOTES
birds	9	Most of the pieces recovered were too fragmentary to allow identification of genera or species. Robinson notes that he was served chicken and eggs during his visit to the mission around 1829 (quoted in Webb 1952:180). Webb also notes that domesticated turkeys, ducks, geese, and doves were raised (1952:187).
canines	1	Members of the Canidae group represent a subgroup of the large land mammal group. The single specimen recovered represents a possible dog (*Canis familiaris*) or coyote (*Canis lestes*) canine tooth. In 1822, Peyri noted that dogs were kept at Mission San Luis Rey. These animals were used to provide security, manage sheep, and drive pests from the fields (Engelhardt 1922:4; Webb 1952:185–86).
cattle	11	*Bos taurus* elements represent a subgroup of the large land mammal group. The specimens recovered include teeth and vertebrae. Many of the scapula, ribs, and teeth fragments counted among the unidentified large land mammal bones probably represent this species. In 1822, Peyri recorded that cattle were slaughtered each Saturday at Mission San Luis Rey to provide food for the neophytes (Engelhardt 1921:41). Webb describes the races of cattle found in the California missions (1952:174–176).
fish	6	The remains of members of the class *Osteichthyes* (Fish bones) are generally small, have a peculiar form, and are distinct from those of terrestrial creatures. Most fish bones have a distinctive rough texture. The head is made up of many delicate plates that are unlikely to be preserved in a recognizable arrangement. The most common fish bone recovered is the vertebrae, which are wasted and have two concave ends. In 1822, Peyri noted that fish were still being collected by the Indians at Mission San Luis Rey from nearby seashores (Engelhardt 1922:41–42).
deer	1	Deer (*Odecoileus* sp.) represents a subcategory of terrestrial land mammals. A single specimen of a long bone was recovered.
horses (and similar species)	1	Equidae bones (Burro, horse, and mule remains (*Equus* spp.) are difficult to distinguish in small fragments. Many of the fragments of post cranial elements are similar to those of cattle. This group represents a subcategory of large terrestrial land mammals. By 1822, the mission herds included 500 horses an 150 mules (Engelhardt 1921:41).
rodents	7	Rodentia (rodent) bones are generally fairly small and hollow. Although some pieces will be so fragmentary that they will not be recognizable, most bones from these creatures can be identified as to genera.
sheep and goats	12	Sheep (*Ovis aries*) and goat (*Capra hircus*) bones are strikingly similar and are referred to as caprine. The elements recovered from the Sunken Gardens include vertebrae, rib, and hoof. This group represents a subcategory of large terrestrial land mammals. By 1822, the mission herds included 20,500 sheep (Engelhardt 1921:41).
unidentified large land mammals	405	Large land mammal bones are usually comparatively massive in size and are generally not hollow. Most of the large land mammal bones found on California colonial sites represent domesticated animals.
unidentified sea mammals	4	Sea mammal remains are similar to large terrestrial land mammal bones except that the bones have a peculiar interior bone structure.
unidentified small mammals	4	Small land mammal bones are similar to, but smaller than those of the large land mammal category. This category includes mammals smaller than an adult cat. Some bones from immature larger mammals will also fall into this category.
unidentified bones	29	This category includes very small fragments that cannot be otherwise categorized.
TOTAL	**490**	

Table 5. *Summary of types of marine shell recovered from the sunken gardens, 2003.*

SPECIES	#	NOTES
Mytilus sp. (mussel)	7	The specimens recovered possibly represent *Mytilus californianus* and *Mytilus edulis*; these types generally inhabit rocky coastal waters; *Mytilus* was used primarily as a food source
Tivela stultorum (Pismo clam)	9	Generally inhabit sandy or muddy waters up to a depth of 80 feet (24 meters); used as a food source and to manufacture jewelry
Pecten sp. (scallops)	13	The specimens present possibly represent *Argopecten circularis* (Pacific Calico Scallops) or *Pecten diegensis* (San Diego Scallops); *Pecten* sp. generally inhabit sandy or muddy waters; used primarily as a food source
Chione sp. (clams)	41	Possibly *Chione californiensis* (California Venus shells), or *Chione undatella*; generally inhabit sandy or muddy waters up to 150 feet deep (18–46 meters); used primarily as a food source
Haliotis sp. (abalone)	4	Includes Green Abalone (*Haliotis fulgens*), red abalone (*Haliotis rufescens*), and black abalone (*Haliotis cracherodii*). Available along rocky seashore, with waters typically 10 to 25 feet deep (3–7.6 meters); used for jewelry and food.
Donax californica (bean clams)	6	Available in local coastal waters in sandy beaches; generally inhabit sandy or muddy waters; used primarily as a food source.
Olivela biplicata (Olivela)	1	A single bead made from this shell was recovered. De Soto noted "hundreds" of "Indian" beads were recovered during the restoration and related excavation projects (1961:34); used exclusively for jewelry.
Ostrea sp. (oyster)	1	Probably *Ostrea lurida*; used primarily as a food source.
Unidentified gastropod	1	Possibly *Natica clausa*; used primarily as a food source.
Unidentified marine shell	186	This category includes small, or highly eroded bleached fragments. Most probably represent fragments of some kind of clam.
TOTAL	269	

Table 6. *Artifact pattern — sunken gardens, 2003.*

ARTIFACT TYPE	COUNT	PERCENT	NOTES	
building materials	5118	52.3%	construction related materials including Roman Cement and tile (tejas and ladrillos)	
ceramics	3836	39.2%	82.3%	trade, food preparation, storage, and serving
glass	13	0.1%	0.3%	
shell	269	2.8%	5.8%	shellfish gathering, jewelry manufacture; gathering, food preparation; and consumption
bone	490	5.0%	10.5%	hunting, fishing, food preparation, and consumption
chipped and ground stone	44	0.4%	0.9%	hunting, food preparation, cutting, abrading, manufacturing, grinding, scraping
limestone	8	0.1%	0.2%	quick lime production
other	77*	*	*	includes organic materials, and miscellaneous other materials
TOTAL	9778	100%	100%	The second column presents percentages without the inclusion of building materials.

Note: only artifacts that can be assigned to the colonial period have been included. A total of 1133 objects were assigned to the modern period.
* Other materials excluded from total.

ILLUSTRATIONS

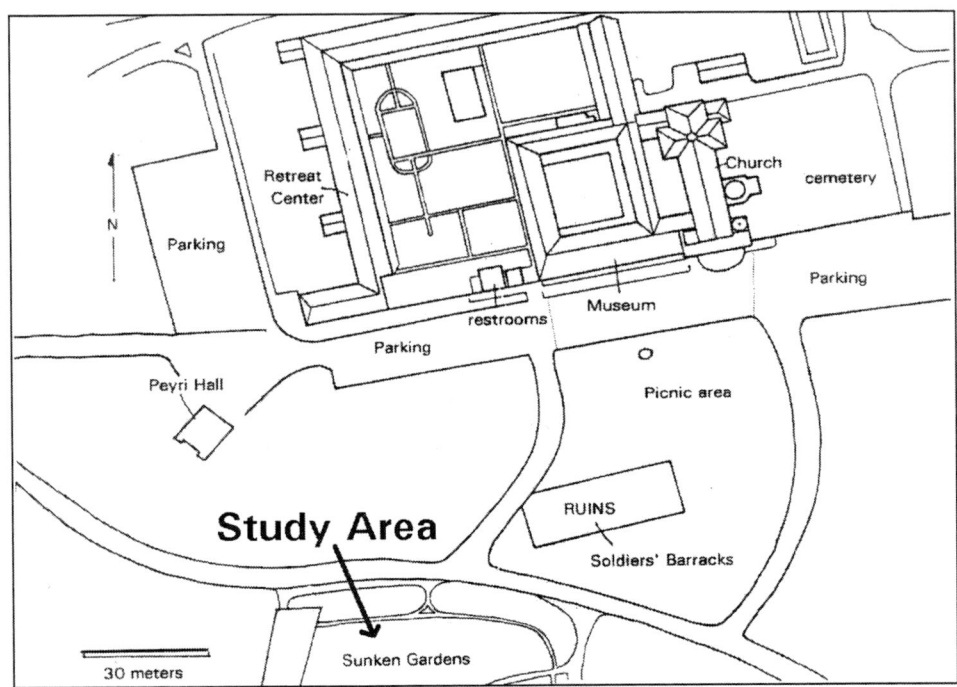

Figure 1. *Map of the site location. Mission San Luis Rey.*

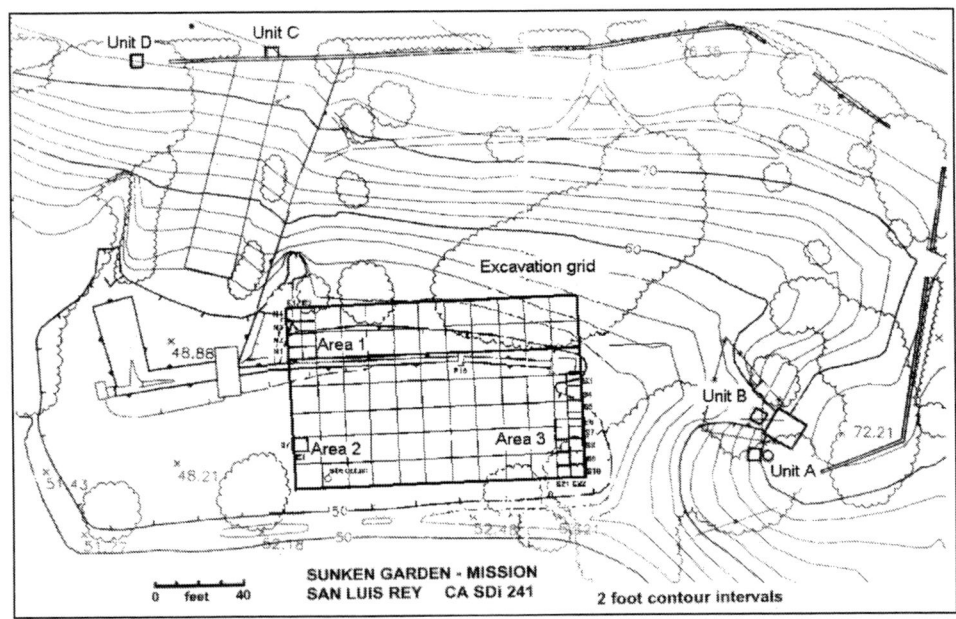

Figure 2. *Map of the site area. Sunken Gardens, Mission San Luis Rey.*

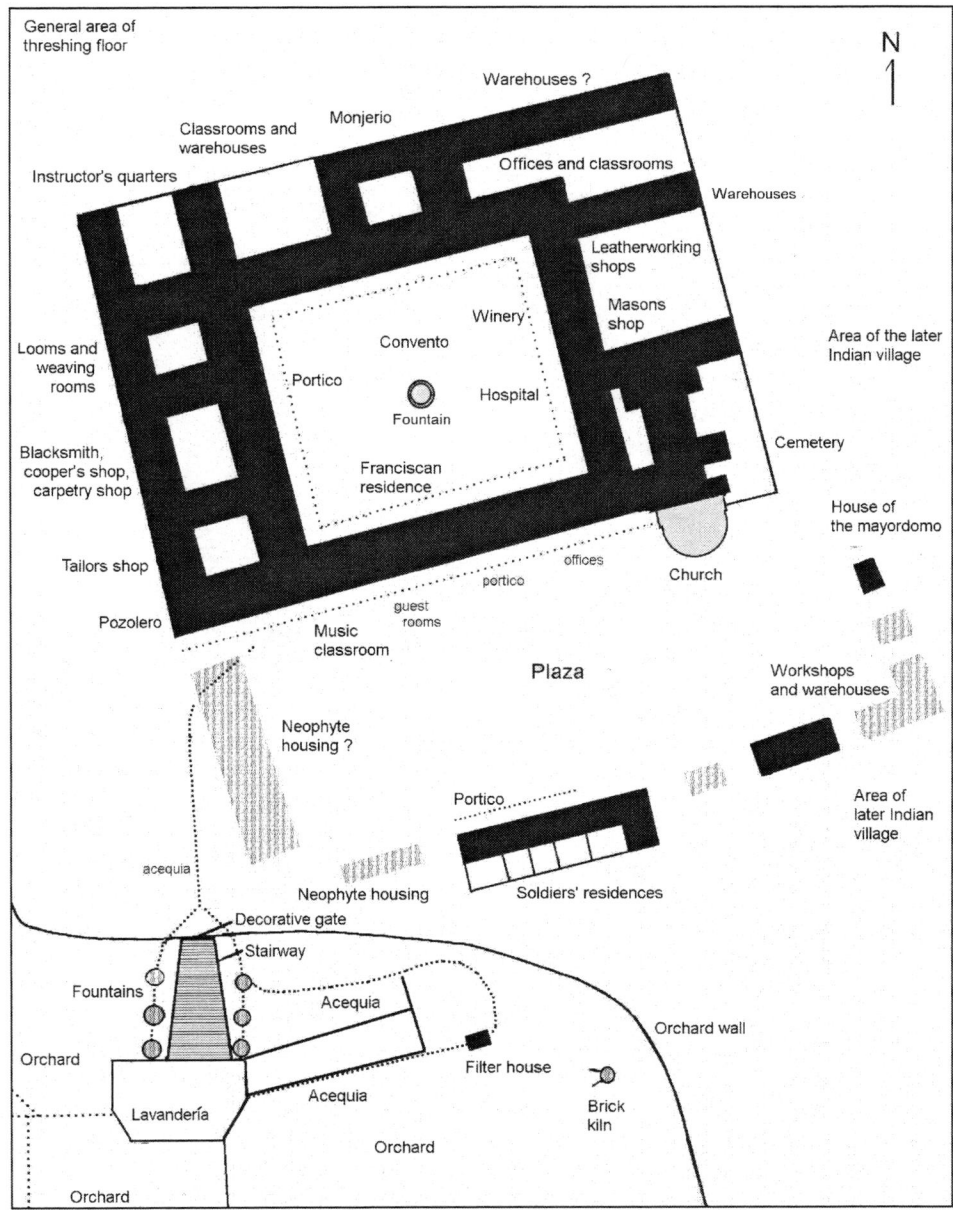

Figure 3. *Plan of Mission San Luis Rey, latter part of Mission Era, c. 1830.*

WAITING FOR TSUNAMI

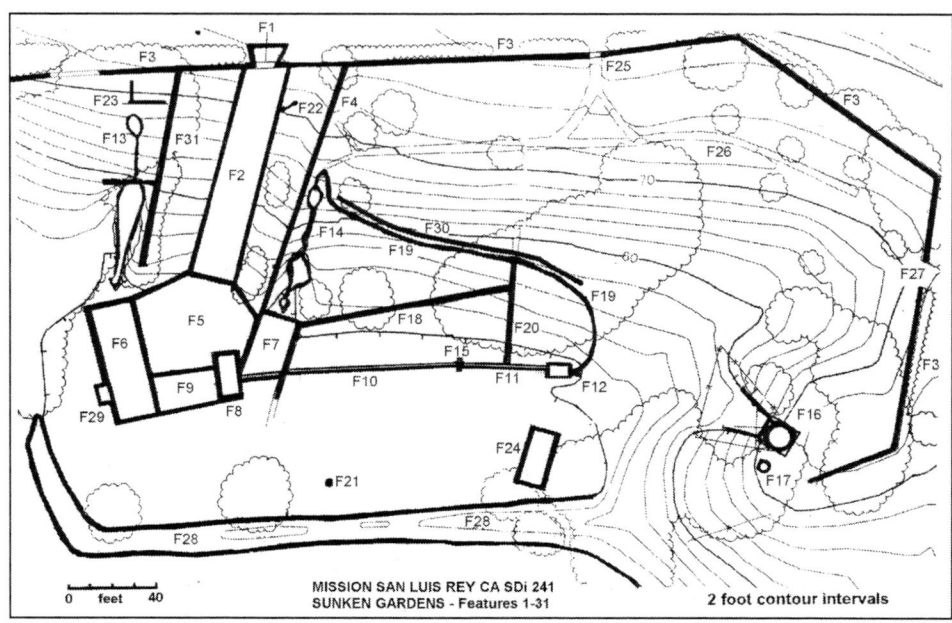

Figure 4. *Features 1 to 31, Sunken Gardens. Mission San Luis Rey.*

Figure 5. *Feature 1: the restored appearance of the monumental gateway from the north.*

THE SUNKEN GARDENS OF MISSION SAN LUIS REY, CALIFORNIA

Figure 6. *Feature 2: the restored appearance of the stairway that drops into the gardens (from the south).*

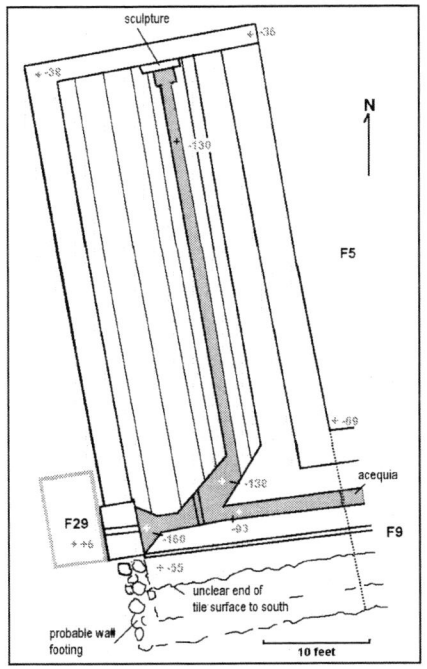

Figure 7. *Plan view of feature 6, the western lavandería.*

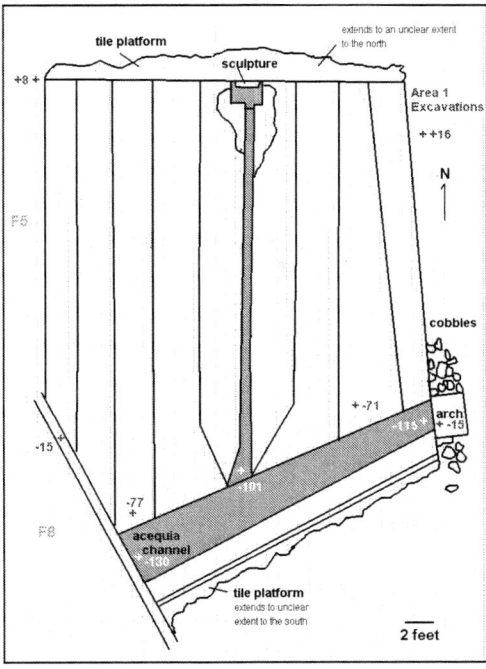

Figure 8. *Plan view of feature 7, the eastern lavandería.*

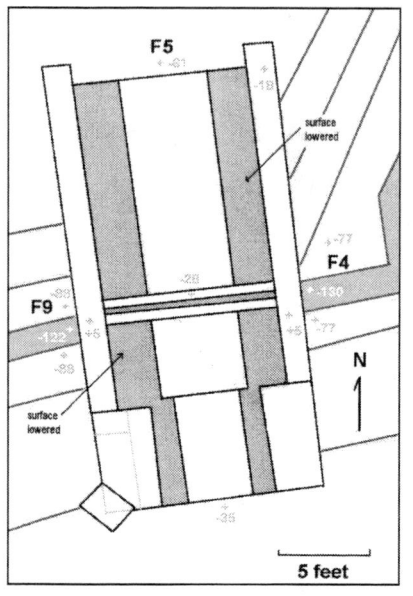

Figure 9. *Plan of feature 8, bridge.*

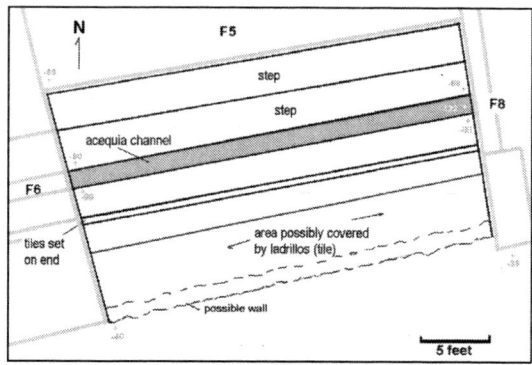

Figure 10. *Plan of acequia segment connecting feature 8 and feature 6.*

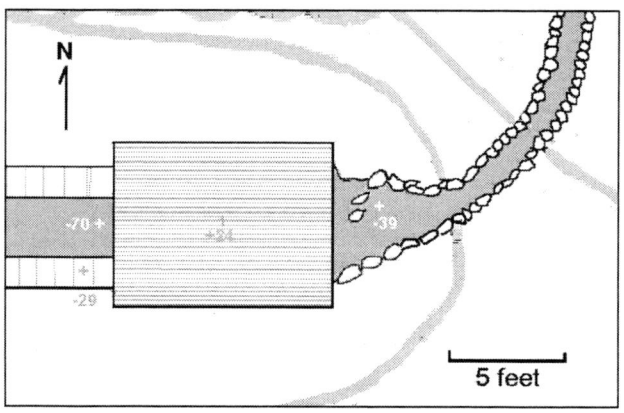

Figure 11. *Feature 12, filter house, plan.*

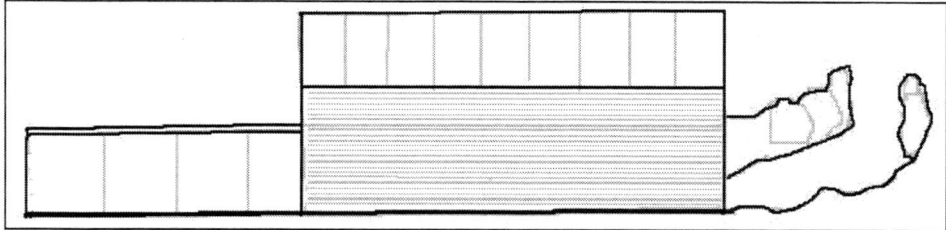

Figure 12. *Feature 12, filter house, (cutaway).*

THE SUNKEN GARDENS OF MISSION SAN LUIS REY, CALIFORNIA

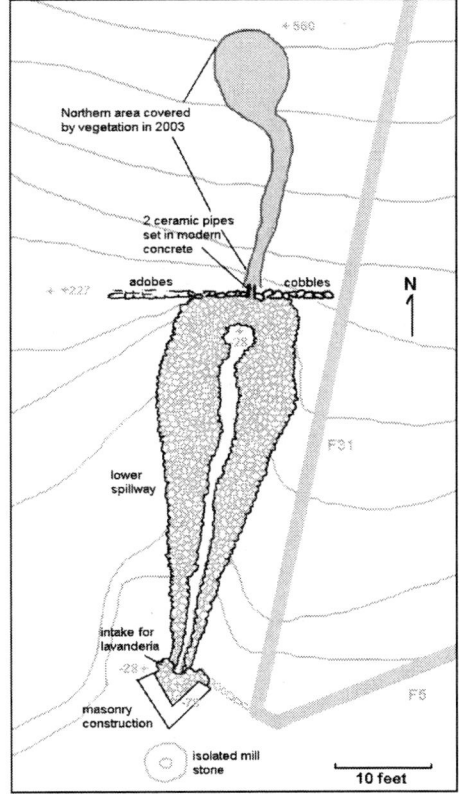

Figure 13. *Feature 13, western fountains, plan.*

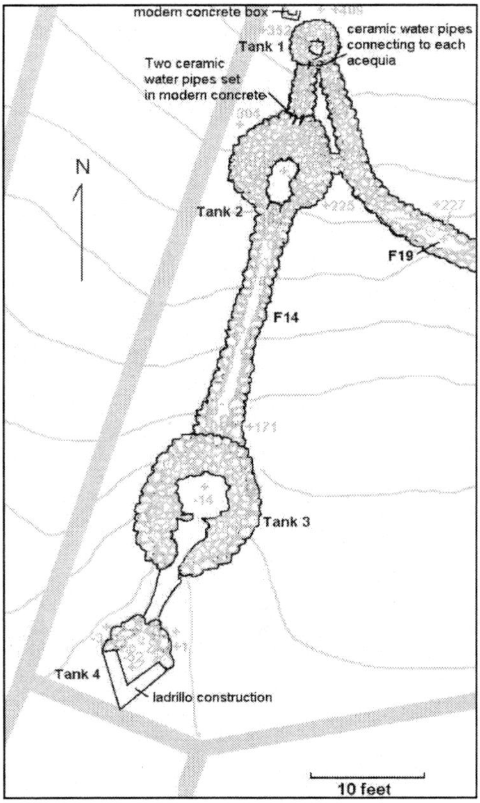

Figure 14. *Feature 14. Eastern fountains, plan.*

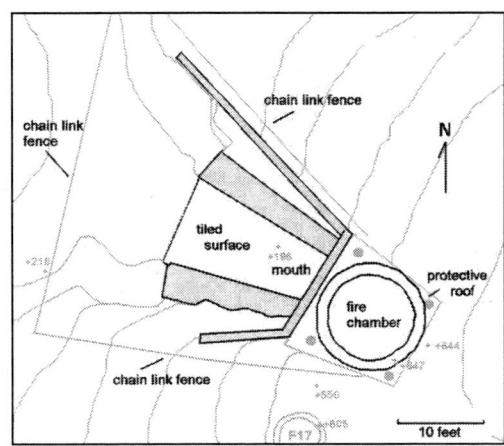

Figure 15. *Feature 16, large kiln, plan.*

Figure 16. *Feature 16, large kiln, reconstruction of throat work area.*

Figure 17. *Feature 17, small kiln, reconstruction (exterior).*

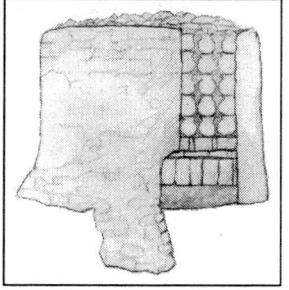

Figure 18. *Feature 17, small kiln, reconstruction (cutaway).*

Figure 19. *Feature 24, tanning vats, plan.*

THE SUNKEN GARDENS OF MISSION SAN LUIS REY, CALIFORNIA

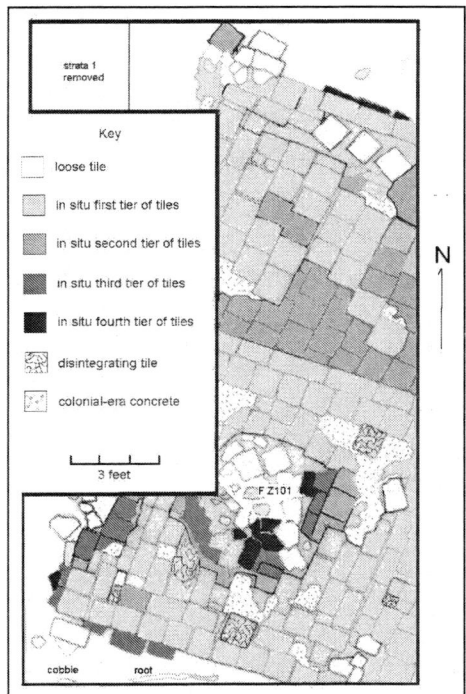

Figure 20. *Feature 24, tanning vat excavations, plan.*

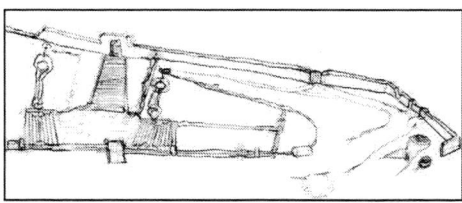

Figure 21. *The Features of the Sunken Garden as they exist today from the south.*

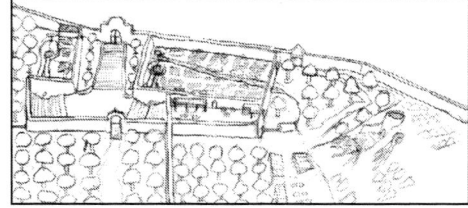

Figure 22. *The Features of the Sunken Garden as they existed in 1830 from the south.*

Figure 23. *The Study area from the west as it may have appeared about 1830.*

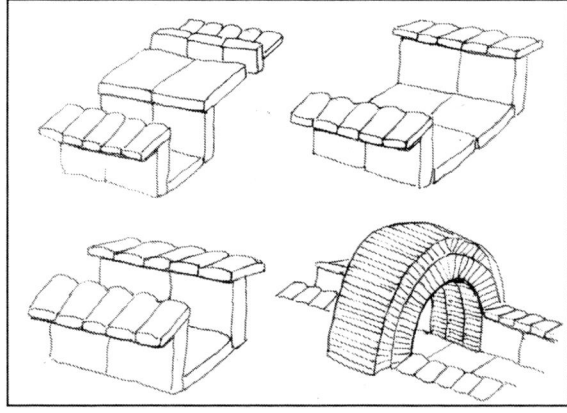

Figure 24. *Various ladrillo construction arrangements used in the acequias of the Sunken Gardens.*

263

REFERENCES

Allen, Rebecca, 1998, *Native Americans at Mission Santa Cruz, 1791-1834, Interpreting the Archaeological Record,* Institute of Archeology, University of California, Los Angeles.

Anonymous, n.d., Artists View (of Mission San Luis Rey), reproduced in Magalousis and Kelsey 1992:101.

Anonymous, 1865, Ground Plan of San Luis Rey Mission according to the United States Survey, reproduced in Engelhardt 1921:236.

Bancroft, Hubert Howe, 1884, *History of California* (volume 1. 1542-1800), A. L. Bancroft & Company, San Francisco.

Barrett, Elinore M., 1987, *The Mexican Colonial Copper Industry,* University of New Mexico Press, Albuquerque.

Bartlett, John R., 1852, Woodcut of Mission San Luis Rey, reproduced in Hart (1967:66).

Boyle, Maida, 1968, *San Luis Rey Mission: Report on the Historical and Archaeological Study,* 1968, reproduced by Magalousis and Kelsey (1991:125-149).

Cheek, Annetta, 1974, *The Evidence for Acculturation in Artifacts: Indians and Non-Indians at Mission San Xavier del Bac, Arizona,* Unpublished doctoral dissertation, University of Arizona, Tucson.

Cheever, Dayle Marie, 1982, An Analysis of the Macrofauna from the Royal Presidio of San Diego Site. Unpublished MA thesis, Department of Anthropology, San Diego State University.

Cleal, John G., 1854, Plat of Mission San Luis Rey prepared for Bishop Alemany, San Luis Rey Mission archive.

Costello, Julia G., 1977, Lime Processing in Spanish California with Special Reference to Santa Barbara. *Pacific Coast Archaeological Society Quarterly* 13:3:22-32.

Costello, Julia G., 1989, *Santa Inés Mission excavations: 1986-1988,* Coyote Press, Salinas.

Deetz, James, 1962, Archaeological investigations at La Purísima Mission, *Archaeological Survey, Annual Report, Department of Anthropology and Sociology, University of California at Los Angeles,* volume 5:161-244.

Di Peso, Charles, 1974, *Casas Grandes: A Fallen Trading Center of the Gran Chichimeca,* Amerind Foundation, Dragoon.

Duby, Georges, 1988, Communal Living, in *A History of Private Life (Volume 2: Revelation of the Medieval World),* edited by Georges Duby, Beknap Press of Harvard University, Cambridge, pp. 35-85.

Duell, Prentice, 1921, Plan to Mission San Luis Rey, reproduced in Engelhardt 1921:68.

Dufloat du Mofras, 1845, *Plan Geometrique de la Mission De St. Louis Roi de France dans la Nouvelle Californie*, reproduced in Egenhoff 1952:56. This map was drafted during an 1841 visit.

Duhaut-Cilly, Auguste Bernard, 1829, Mission San Luis Rey (woodcut) reproduced in Duhaut-Cilly 1834.

Duhaut-Cilly, Auguste Bernard, 1834, *Voyage Autour du Monde Principlamente a la California* (2 volumes) A. Bertrand, Paris.

Duhaut Cilly, Auguste Bernard (translated by Charles Franklyn Carter), 1929, Duhaut-Cilly's Account of California in the Years 1827-1828 [originally published in 1834 as *Voyage Autour du Monde*, Saint-Servan, Paris]. *Quarterly of the California Historical Society* volume 8(2-June):130–167.

Egenhoff, Elizabeth (editor), 1952, *Fabricas: A collection of pictures and statements on the Mineral materials used in buildings in California*. State of California, Sacramento.

Engelhardt, Zephyrin (Fray), 1920, *San Diego Mission*, James H. Barry Company, San Francisco.

Engelhardt, Zephyrin (Fray), 1921, *San Luis Rey, the King of the Missions*, James H. Barry Company, San Francisco.

Farris, Glenn, 1991, Archaeological Testing in the Neophyte Family Housing Area at Mission San Juan Bautista, California. Manuscript on file, Archaeology Lab, Department of Parks and Recreation, West Sacramento.

Farris, Glenn, 1995, Archaeological Evaluation of the Neophyte Family Housing (plus Infirmary), the Granary, and the Threshing Floor at La Purísima Mission State Historic Park, manuscript on file, Archaeology Lab, Department of Parks and Recreation, West Sacramento.

Greenwood, Roberta, 1975, *3500 years on a city block: San Buenaventura Mission Plaza Project archaeology report*, Greenwood and Associates, Pacific Palisades.

Hancock, Henry, 1860, Plat of Four Tracts of Land of the Ex-Mission San Luis, copy on file with the San Luis Rey Mission Archive.

Hart, Herbert M., 1967, *Pioneer Forts of the West*, Superior Publishing Company, Seattle.

Hartnell, William E. P. (translated by Starr Pait Gurcke, and edited by Glenn J. Farris), 2004, *The Diary and Copybook of William E. P. Hartnell, Visitador General of the Missions of Alta California in 1839 and 1840*, California Mission Studies Association and the Arthur H. Clarke Company, Santa Clara, Spokane.

Healy, Valentine (prepared for publication by Thomas Davis), 1992, *Father O'Keefe Rebuilder of Mission San Luis Rey,*, Padre Press, Northridge, originally published in Times Gone By, volume 11, 1965:16–25.

Henderson, Richard R. (editor), 1989, *A Preliminary Inventory of Spanish Colonial Resources associated with National Park Service Units and National Park Landmarks, 1987 (second edition)*, United States Department of the Interior, Washington.

Hewitt, James M., 1975, The Faunal Archaeology of the Tubac Presidio. In Shenk and Teague 1975:199–233.

Hoover, Robert L. and Julia Costello (editors), 1985a, Excavations at Mission San Antonio 1976–1978, *Institute of Archaeology Monograph 26*, University of California at Los Angeles.

Osio, Antonio María (translated by Rose Marie Beebe and Robert M. Senkewicz), 1996, *The History of Alta California*, University of Wisconsin Press, Madison.

Jackson, Robert H. and Edward Castillo, 1995, *Indians, Franciscans and Spanish Colonization: The Impact of the Mission System on California Indians*, University of New Mexico Press, Albuquerque.

Kelsey, Harry, 1993, *Mission San Luis Rey: A Pocket History*, Interdisciplinary Research Incorporated, Altadena.

Kelsey, Harry and Nicholas Magalousis, 1991, *Archaeological and Historical Investigation at Mission San Luis Rey, California, CA-SDi-241, Sector C, for the Peyri Road Water Line Project (711-85-7850) of the City of Oceanside Water Utilities Department*, Interdisciplinary Research, Laguna Beach.

McCowen, Benjamin E., 1955, Temeku, A Page from the History of the Luiseño Indians, *Archaeological Survey Association of Southern California*, Paper Number 3, Los Angeles.

Magalousis, Nicolas and Harry Kelsey, 1992, *Preliminary Archaeological and Historical Investigations at Mission San Luis Rey, California CA-SDi-241: Sector D:Phase I Report. Volume V: San Luis Rey: Friary and Cemetery*, Interdisciplinary Research, Laguna Beach.

Miller, Henry, 1856, Drawing of Mission San Luis Rey (from the South), in Miller 1985:56.

Miller, Henry, 1985, *Account of a Tour of the California Missions & Towns 1856 The Journal & Drawings of Henry Miller*, Bellerophon Books, Santa Barbara.

Oak, Henry, 1874, Mission San Luis Rey (plan), reproduced in Magalousis and Kelsey 1992:8.

Powell, H. M. T., 1850, Mission San Luis Rey (drawing) reproduced, in Powell 1931:opposite 204.

Powell, H. M. T., 1850a, Military barracks at Mission San Luis Rey, reproduced in Kelsey 1993:22.

Powell, H. M. T., 1931, *The Santa Fe Trail to California, 1842-1846*, Grabhorn Press, San Francisco.

Regnier-Bohler, Danielle, 1988, Imagining the Self, in *A History of Private Life (Volume 2: Revelation of the Medieval World)*, edited by Georges Duby, Beknap Press of Harvard University, Cambridge, pp. 311-394.

Robinson, Alfred, 1829, Woodcut of Mission San Luis Rey, included in Robinson 1846.

Robinson, Alfred, 1846, *Life in California During a Residence of Several Years in that Territory*, Wiley and Putnam, New York, reprinted in 1969 by Da Capo Press, New York.

Romani, John F. and A. George Toren, 1975, A preliminary analysis of faunal remains from the Ven-87 aboriginal and historic components: Phase I. In Greenwood 1975.

Roncière, Charles, 1988, Tuscan Nobles on the Eve of the Renaissance, in *A History of Private Life (Volume 2: Revelation of the Medieval World)*, edited by Georges Duby, Beknap Press of Harvard University, Cambridge, pp.157-310.

Salomon, Carlos, 2004, Pio Pico at Mission San Luis Rey, 1835-1840: A Study in Mission Administration, in The Mission and the Community, *Proceedings of the 21st Annual Conference of the California Mission Studies Association, San Luis Obispo, California February 13-15*, edited by Dan Krieger, California Mission Studies Association, Santa Clara, pp. 86-96.

Schuetz (Schuetz-Miller), Mardith K., 1994, *Buildings and Builders in Hispanic California, 1769-1850*. Southwestern Mission Research Center and the Santa Barbara Trust for Historic Preservation, Tucson, and Santa Barbara.

Shenk, Lynette and George A. Teague, 1975, Excavations at Tubac Presidio, *Arizona State Museum Archaeological Series, no. 85*.

Soto, Anthony, 1960, Recent Excavations at San Luis Rey Mission: The Sunken Gardens, *Provincial Annals, Province of Santa Bárbara*, Order of Friars Minor (April 1960) 22:205-21, 247-49.

Soto, Anthony, 1961, Mission San Luis Rey, California — Excavations in the Sunken Gardens, *The Kiva: A Journal of the Arizona Historical Society* (April 1961) 26:34-43.

Tac, Pablo (translated by Minna and Gordon Hewes), 1952, Indian Life and Customs at Mission San Luis Rey, *The Americas* 9(1):87-106.

Walker, Phillip L. and Katherine D. Davidson, 1989, Analysis of faunal remains from Santa Inés Mission. In Costello 1989:162–176.

Webb, Edith, 1952, *Indian Life at the Old Missions*. Wayside Press, Los Angeles.

Weber, Francis J., 2001, *Encyclopedia of California's Catholic Heritage,* Arthur H. Clarke and Saint Francis Historical Society, Mission Hill and Spokane.

Williams, Jack S., 1995, Architectural Analysis and Recommendations for the Capistrano Kiln Site. A report prepared for Milford Wayne Donaldson, on file with the Center for Spanish Colonial Research Archive, San Diego.

Williams, Jack S., 2004, An Archaeological Investigation of the Sunken Gardens at Mission San Luis Rey, manuscript, Center for Spanish Colonial Research, San Diego.

Williams, Jack S. and Anita G. Cohen-Williams, 1997, *Center for Spanish Colonial Archaeology Handbook,* The Center for Spanish Colonial Archaeology: San Diego.

Mission San Luis Rey, Oceanside, February 1, 1991.
Photo: Woodrow L. Higdon, Geo-Tech Imagery.

View north along the Pacific coastline from San Onofre State Beach Park to San Onofre Nuclear Generating Station. San Onofre State Beach landslide blocks and down-dropped terraces can be observed in the center of the photograph. See Road Log Stop 3 for additional information. Photo: Diane Murbach.